Integrated Computer Technologies in Mechanical Engineering – Synergetic Engineering

Integrated Computer Technologies in Mechanical Engineering – Synergetic Engineering

Editors

Mykola Nechyporuk
Volodymyr Pavlikov
Dmytro Krytskyi

Basel • Beijing • Wuhan • Barcelona • Belgrade • Novi Sad • Cluj • Manchester

Editors

Mykola Nechyporuk
Cars and Transport
Infrastructure
National Aerospace
University Kharkiv Aviation
Institute
Kharkiv
Ukraine

Volodymyr Pavlikov
Aerospace Radioelectronic
Systems
National Aerospace
University Kharkiv Aviation
Institute
Kharkiv
Ukraine

Dmytro Krytskyi
Information Technology
Design
National Aerospace
University Kharkiv Aviation
Institute
Kharkiv
Ukraine

Editorial Office
MDPI
St. Alban-Anlage 66
4052 Basel, Switzerland

This is a reprint of articles from the Special Issue published online in the open access journal *Computation* (ISSN 2079-3197) (available at: www.mdpi.com/journal/computation/special_issues/ICTM2021).

For citation purposes, cite each article independently as indicated on the article page online and as indicated below:

Lastname, A.A.; Lastname, B.B. Article Title. *Journal Name* **Year**, *Volume Number*, Page Range.

ISBN 978-3-7258-0830-4 (Hbk)
ISBN 978-3-7258-0829-8 (PDF)
doi.org/10.3390/books978-3-7258-0829-8

© 2024 by the authors. Articles in this book are Open Access and distributed under the Creative Commons Attribution (CC BY) license. The book as a whole is distributed by MDPI under the terms and conditions of the Creative Commons Attribution-NonCommercial-NoDerivs (CC BY-NC-ND) license.

Contents

About the Editors vii

Preface ix

Vladimir Shevel, Dmitriy Kritskiy and Oleksii Popov
Toward Building a Functional Image of the Design Object in CAD
Reprinted from: *Computation* **2022**, *10*, 134, doi:10.3390/computation10080134 1

Vitaly Miroshnikov, Basheer Younis, Oleksandr Savin and Vladimir Sobol
A Linear Elasticity Theory to Analyze the Stress State of an Infinite Layer with a Cylindrical Cavity under Periodic Load
Reprinted from: *Computation* **2022**, *10*, 160, doi:10.3390/computation10090160 24

Valeriy Volosyuk, Volodymyr Pavlikov, Simeon Zhyla, Eduard Tserne, Oleksii Odokiienko, Andrii Humennyi, et al.
Signal Processing Algorithm for Monopulse Noise Noncoherent Wideband Helicopter Altitude Radar
Reprinted from: *Computation* **2022**, *10*, 150, doi:10.3390/computation10090150 35

Oleksandr Zabolotnyi, Vitalii Zabolotnyi and Nikolay Koshevoy
Capacitive Water-Cut Meter with Robust Near-Linear Transfer Function
Reprinted from: *Computation* **2022**, *10*, 115, doi:10.3390/computation10070115 51

Oleksii Tretiak, Dmitriy Kritskiy, Igor Kobzar, Mariia Arefieva and Viacheslav Nazarenko
The Methods of Three-Dimensional Modeling of the Hydrogenerator Thrust Bearing
Reprinted from: *Computation* **2022**, *10*, 152, doi:10.3390/computation10090152 70

Sergiy Yakovlev, Oleksii Kartashov and Dmytro Podzeha
Mathematical Models and Nonlinear Optimization in Continuous Maximum Coverage Location Problem
Reprinted from: *Computation* **2022**, *10*, 119, doi:10.3390/computation10070119 78

Yurii Pronchakov, Oleksandr Prokhorov and Oleg Fedorovich
Concept of High-Tech Enterprise Development Management in the Context of Digital Transformation
Reprinted from: *Computation* **2022**, *10*, 118, doi:10.3390/computation10070118 93

Dmytro Chumachenko, Ievgen Meniailov, Kseniia Bazilevych, Tetyana Chumachenko and Sergey Yakovlev
Investigation of Statistical Machine Learning Models for COVID-19 Epidemic Process Simulation: Random Forest, K-Nearest Neighbors, Gradient Boosting
Reprinted from: *Computation* **2022**, *10*, 86, doi:10.3390/computation10060086 112

Ihor Romanenko, Yevhen Martseniuk and Oleksandr Bilohub
Modeling the Meshing Procedure of the External Gear Fuel Pump Using a CFD Tool
Reprinted from: *Computation* **2022**, *10*, 114, doi:10.3390/computation10070114 134

Nutthapong Kunla, Thira Jearsiripongkul, Suraparb Keawsawasvong and Chanachai Thongchom
Crack Identification in Cantilever Beam under Moving Load Using Change in Curvature Shapes
Reprinted from: *Computation* **2022**, *10*, 101, doi:10.3390/computation10060101 153

About the Editors

Mykola Nechyporuk

Mykola Nechyporuk was born on November 10, 1952, in a village in the Malevo Rivne region. In 1973, he came to the Kharkov Aviation Institute at the Radio Engineering Faculty to study the design and manufacture of radio equipment. After graduation in 1979, he worked as an engineer at the Kharkov Aviation Institute. In the same year, he became the director of the KhAI campus.

Since 1998, M. Nechyporuk has been vice-rector for administrative and financial activities at the KhAI. From 1998 to 2010, he worked as a deputy of the Kiev District Council. From 2010 to 2015, he was a deputy of the Kharkiv City Council. From 2004 to 2018, he served as vice-rector for scientific and pedagogical work at the National Aerospace University, Kharkiv Aviation Institute. In his practice, he has conducted a number of scientific studies, such as problems of integrated technologies for the assembly and utilization of aircraft and other vehicles; technologies for the implementation of protective coatings for aircraft and automotive vehicles; ontology modeling; and ontological decision support systems for the choice of impulse devices. He developed a system for supporting decision-making in the activities of a university, on the basis of which, for the first time, a computer-based decision support system was introduced among universities in Ukraine.

He is the rector of the National Aerospace University, Kharkiv Aviation Institute, 2018–2023. He has more than a hundred scientific papers, including monographs, textbooks, publications in foreign journals, and publications in world scientometric databases. His achievements include the following: Doctor of Technical Sciences (2012), Professor (2011), Mark of Excellence in Education of Ukraine (2002), Honored Educator of Ukraine (2005), and being awarded with the Diploma of the Verkhovna Rada of Ukraine for merits to the Ukrainian people (2013).

Volodymyr Pavlikov

Vladimir Pavlikov was the Dean of the Radio Electronics, Computer Systems, and Info-Communications Faculty (April 2017–September 2018). He was head of the department of aircraft radio-electronic systems design (April 2015–March 2017). He has received an academic degree, a Doctor of Technical Sciences (since January 2014). He has a scientific rank of senior researcher (since April 2015).

In 2014, he was awarded the Prize of the Verkhovna Rada of Ukraine for young scientists for a series of scientific works that developed the theory of the synthesis of wideband radiometric systems. In 2015, he won the First Prize (European Microwave Association). In the period 2013–2018, he was a scholar of the Cabinet of Ministers of Ukraine for young scientists. He is a reviewer of the following scientific journals: *IEEE Access*, *Physical Bases of Instrumentation*, and *Radioelectronic and Computer Systems*.

In the period 2016–2018, he was the head of the scientific work titled "New principles of processing signals of own radio-thermal radiation of objects of different physical nature and technologies for their implementation". This project is one of 79 that were selected from more than 400 works recognized as important by the Ministry of Education and Science of Ukraine and assessed by Ukrainian and European scholars for relevance.

From 2019 to 2021, he was the head of the scientific work titled "The development of the theory of ultra-wideband active aperture synthesis systems for high-precision remote sensing from high-speed aerospace platforms".

Dmytro Krytskyi

Dmitriy Kritskiy has a Ph.D. in project and program management. In 2002–2008, he received a specialist degree in information technology from the National Aerospace University, Kharkiv Aviation Institute. He is the Dean of the Faculty of Aircraft Engineering. He was Associate Professor at the National Aerospace University Kharkiv Aviation Institute (2018–2021) and Senior Lecturer at the National Aerospace University Kharkiv Aviation Institute (2015–2017). He was an assistant at the National Aerospace University Kharkiv Aviation Institute in 2014 and a Ph.D. student at the National Aerospace University Kharkiv Aviation Institute in 2010–2013. He was a junior researcher at the National Aerospace University Kharkiv Aviation Institute in 2008–2009. His current position is as Dean of the Aircraft Building Faculty and lecturer on decision-making theory, information system design, digital information protection, and virtual reality technology. His research interests include developing software for unmanned aerial vehicles (UAVs) and automatic control systems for multi-copter group fights. He has 100 peer-reviewed papers and book chapters, participated in nine research projects, four of which he headed, and has 15 patents.

Preface

The best works were selected after the conference titled "Integrated Computer Technologies in Mechanical Engineering"– Synergetic Engineering (ICTM).

The conference provided technical exchange between the scientific community in the form of keynote addresses, panel discussions, and a special session. In addition, participants were treated to a series of techniques that facilitated collaboration among fellow researchers. ICTM'2021 received 110 applications from different countries.

All this gives us a lot of valuable information and will greatly benefit the exchange of experience between scientists in the field of modeling and simulation. The organizers of ICTM 2021 have made great efforts to ensure the success of this conference. We would hereby like to thank all members of the ICTM 2021 Advisory Committee for their guidance and advice; members of the program committee and organizing committee; reviewers for their efforts in reviewing and collecting articles; and all authors for their contributions to creating a unified intellectual environment for solving current scientific problems.

We express our gratitude to the MDPI publishing house and the editors of *Computation* Journal for the opportunity to publish the best research.

Mykola Nechyporuk, Volodymyr Pavlikov, and Dmytro Krytskyi
Editors

Article

Toward Building a Functional Image of the Design Object in CAD

Vladimir Shevel, Dmitriy Kritskiy * and Oleksii Popov

Department of Information Technology Design, National Aerospace University "Kharkiv Aviation Institute", 61070 Kharkiv, Ukraine
* Correspondence: d.krickiy@khai.edu

Abstract: The paper proposes an approach to the classification of lifecycle support automation systems for engineering objects, with the proposed structure of the description of the designed object, using a triple description approach: functional, mathematical, and physical. Following this approach, an algorithm for drawing up a functional description of the lifecycle is described, which is based on the principle of unity of analysis and synthesis of the created system in the design process. The proposed solutions are considered using the traditional aircraft shaping methodology with the application of the airplane make-up algorithm as an example. Furthermore, the architecture of a multiagent platform for structural–parametric synthesis of the object was presented; for convenient usage of this architecture, it was proposed to use classification of design tasks in the form of a design cube. The proposed approach allows obtaining an accurate description of the designed object and the subtasks needed to create it, which can reduce the time of the project. Unfortunately, not all decisions can be automated at the given stage of technical development, but what is possible to automate is enough to achieve a reduction in terms of realization and an acceleration of the prototyping process, as shown in the considered example. The actual reduction throughout the lifecycle of the product ranges from 10% to 21% of the planned time.

Keywords: lifecycle; CAD; engineering design; multiagent platform

Citation: Shevel, V.; Kritskiy, D.; Popov, O. Toward Building a Functional Image of the Design Object in CAD. *Computation* **2022**, *10*, 134. https://doi.org/10.3390/computation10080134

Academic Editors: Demos T. Tsahalis and Anna T. Lawniczak

Received: 20 June 2022
Accepted: 3 August 2022
Published: 5 August 2022

Publisher's Note: MDPI stays neutral with regard to jurisdictional claims in published maps and institutional affiliations.

Copyright: © 2022 by the authors. Licensee MDPI, Basel, Switzerland. This article is an open access article distributed under the terms and conditions of the Creative Commons Attribution (CC BY) license (https:// creativecommons.org/licenses/by/ 4.0/).

1. Introduction

The desire to automate as fully as possible the support of the lifecycle of an engineering object (LCEO) has led to the creation of numerous examples of automated systems in the form of application packages and variants of their combination in the form of integrated computer systems (ICS). Most publications on the subject of lifecycle management are aimed at proving that the formalization of the design process is necessary, in order to achieve a reduction in both implementation time and cost. For example, the authors of [1] showed the need to create a formalized method for obtaining comparable and transparent assessments of lifecycle sustainability indicators in the early stages of design and planning of a civil works project, which allows comparing the sustainability indicators of design concepts at the lifecycle stage and at the level of building components.

The latter position themselves as some "PLM solutions" of a comprehensive nature, built using continuous acquisition and lifecycle support (CALS) technologies on the basis of the unified information space (UIS) of the LCEO. For example, in [2], the LCEO was described using the example of aviation technology to achieve the minimization of harmful emissions throughout the entire lifecycle, while describing several tools that complement each other, but duplicating functionality was also shown.

The authors of [3] demonstrated the diversity of computer-aided design (CAD) systems in modeling, and they also proposed the function and geometry exploration (FGE) approach, which allows designing on the basis of the description of functionality and geometry. As noted in the publication, this is very good for prototyping (obtaining a large number of different options); however, in general, it is not enough for the implementation of lifecycle management at all stages.

The main disadvantage of such tools is the duplication of a number of functions.

The authors of [4] classified and analyzed various lifecycle support models, revealing that existing systems complement, duplicate, and, in some cases, have redundant functionality. These models focus on a continuous development strategy, where the information gathered during the cycles can serve as useful input for managing future projects or even expanding and improving the current project, while taking into account possible risks.

In particular, practically all systems of the considered purpose support elements of design automation with respect to various engineering objects. This is due to the fact that design elements take place when solving almost all tasks of supporting the lifecycle of an engineering object (LCEO). Examples of such systems are the following [5]:

CAD (computer-aided design). They automate the procedures of geometric modeling and design documentation generation in an engineering object design process. A set of geometric models and design documentation is considered as a design object.

CAE (computer-aided engineering). These are automation systems for engineering analysis, which is seen as an element of engineering design. Modern CAE systems are used together with CAD systems, often integrated into them. Physical and mathematical models of the analyzed engineering object are considered as a design object.

CAPP (computer-aided process planning). The object of this design is the technological processes of the production of objects.

CAM (computer-aided manufacturing). Systems for the computer-aided preparation of control programs automate the procedure of control program design. The object of the design is considered to be the software controlling the technological equipment.

CASE (computer-aided system/software engineering). This is a system for automating the development of information systems or software. Software components are considered as the object of design.

CAQ (computer-aided quality control). The object of computer-aided quality control systems is a quality control system for manufactured products.

FMS (flexible manufacturing system). The object of the design is equipment configurations adapted to the production of specific products and their volume.

SCADA (supervisory control and data acquisition). Data acquisition technology for supervisory control of a process is considered as the design object.

CIM (computer-integrated manufacturing). The object of the design is the combination of different production facilities to produce a particular product.

MRP (manufacturing resource planning). The object of the design is a system of plans used in the organization of production.

MRO (maintenance repair and overhaul). The system of documents regulating the rules of operation and maintenance of facilities is considered as the object of design.

ICS ProEngineer (WildFire), Unigrapphics (NX), and CATIA are integrated computer systems to support LCEO, incorporating the functions of the above packages.

The list of examples could go on.

2. Materials and Methods

In this situation, it seems relevant to clarify the concept and content of design, since they are presently individual in nature and depend on the tastes of the developers of automation systems. The solution of this problem makes it possible to unify the approach to automating the design procedure, which creates prerequisites for reducing the duplication of design automation tools in the components of the automated LCEO support systems and in the development of ICS.

The basis for defining the inherent functions of any automation system to be created is to analyze the automation object. In this paper, an attempt is made to identify the functions inherent in CAD system for an engineering object.

In order to solve this problem, it is first necessary to define the concept of design.

Design is one of the most general and "fuzzy" concepts used in engineering practice. The analysis of normative documents related to design [6] does not give an unambiguous answer to this question. Most often, design is indirectly defined through concepts such as "design decision", "design document", "design operation", and "design procedure". In our opinion, the most constructive approach is to define design as a procedure for constructing a complete description of the engineering object to be created.

It is proposed to present a complete description of an engineering object in the form of three components (Figure 1): functional description, mathematical description, and physical description, corresponding to the notions of description of complex objects (systems) contained in [7],

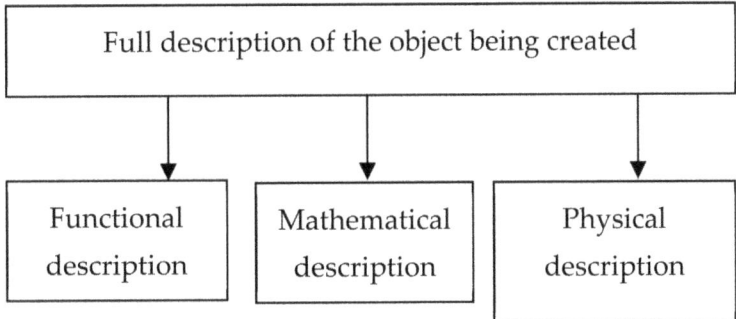

Figure 1. Structure of the complete description of the project.

The functional description or F-image of the object to be created is a necessary and sufficient set of functions performed by the object in its functioning as a system of purposeful action. The basis for selecting a set of functions is the general (integral) function of the designed object. It is subjected to successive analysis (detailed elaboration and splitting) in the design process.

The mathematical (algorithmic) description or M-image of the created object is the result of algorithmization of implementation of F-image components. Depending on the specific algorithms, the functions can be implemented in automatic mode using hardware or software, in automated mode, when human participation in their implementation is required, and in "manual" mode, when the allocated function is performed directly by a human.

The physical (real-life) description or N-image is a description of the structural elements of the designed object intended for the in situ realization of the F-image components, functioning in accordance with the M-image components.

Figure 2 shows a simple example of how to describe the design object. Example 1 is a bridge truss element, and Example 2 is a numerical integration program.

The design procedure as a process of constructing F-, M-, and N-images of an engineering object corresponds to the principle of unity of analysis and synthesis in the course of solving a complex problem (Figure 3).

The design procedure is presented as a sequence of procedures for analyzing the functional purpose of an object, formalizing the resulting set of functions, naturalizing the functions as elements of the design of the object, and then synthesizing the design as a whole.

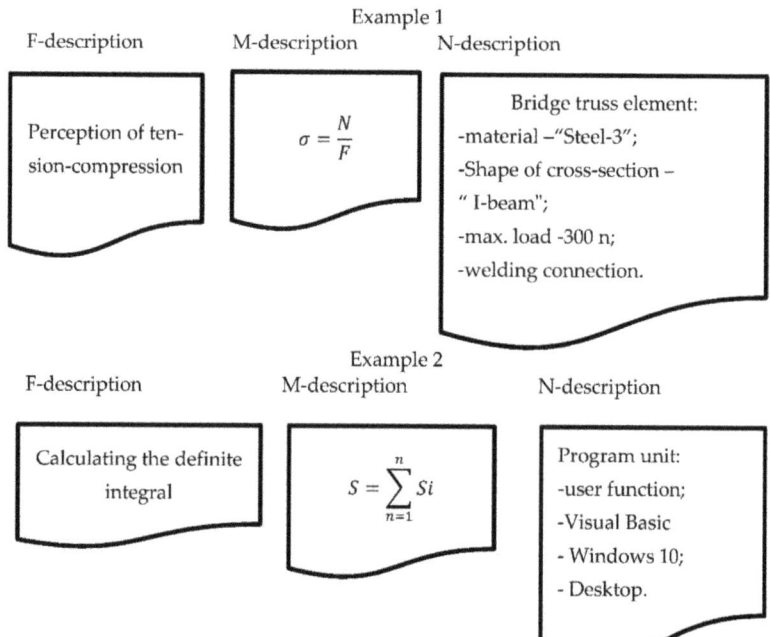

Figure 2. Examples of a complete description.

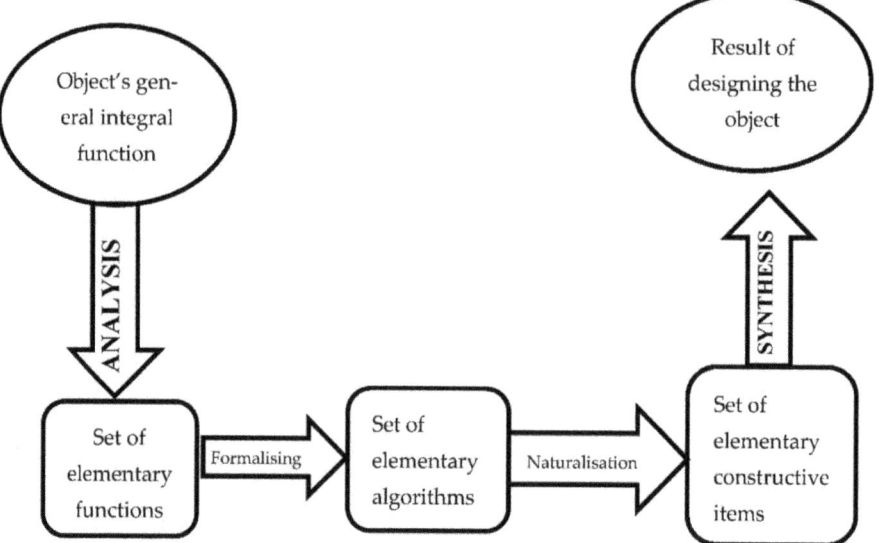

Figure 3. Design as a unity of analysis and synthesis.

Naturally, the proposed scheme may contain feedback, indicating the iterative nature of the design process. Iterations are possible at any stage. The following specifics can be considered:

1. The extracted function cannot be formalized to the required level, and a correction of the overall function analysis is necessary.
2. The resulting mathematical description does not provide a solution to the mathematical problem and needs to be simplified.
3. The resulting algorithm cannot be physically realized within known physical principles and requires correction of the algorithm or additional research to develop new physical principles.
4. The resulting set of design elements cannot be combined into a coherent whole due to existing design or technological constraints.

The number of rollback steps can be more than one.

The specifics of constructing the leading component of the description of the projected object are considered the F-description.

The basis for initiating the design is the summary of the requirements for the design object to be created and its overall (integral) function, which are the result of the predesign procedures.

The result of its subsequent analysis, which determines the completeness of the F-image of the designed object, depends on the correct formulation of the general function. The general function is formulated on the basis of a thorough analysis of the requirements statement for the object to be created. It is beyond the scope of this paper to formulate the set of requirements at the level of pre-project procedures. It should only be noted that the procedure for creating the requirements statement is well developed and can be automated. This is facilitated by the existence of standards for requirements statements, e.g., ISO/IEC/IEEE 29148:2011 for software development. Approaches have been developed and software templates exist to automate the development of a requirements statement.

The correctness of the formulation of the general function can be controlled by having a formal description of it. For example, in [8], it was proposed to use the notion of the main utility function of the system P in the form of

$$P = (D, G, H), \qquad (1)$$

where D is the action carried out in the execution of the function, G is the object on which the action is being performed, and H is the condition under which the action takes place.

The proposed description can be extended by specifying the nature of the data transformation performed during the action. For this purpose, for example, in [9], the notion of a technical function of the system F was introduced as

$$F = (P, Q), \qquad (2)$$

where

$$Q = I \rightarrow O. \qquad (3)$$

In this case, Q defines the nature of the transformation of some set I consisting of n input operands into a set O consisting of m output operands.

$$\{I_1, I_2, I_3, \ldots, I_n\} \rightarrow \{O_1, O_2, O_3, \ldots, O_m\}. \qquad (4)$$

Such a refinement opens up the possibility of formally controlling the correctness of the F-image of the designed object.

The process of analyzing a general function can be represented as its sequential "splitting" of the function F into subfunctions (Figure 4). The authors of [10] showed how many combinations of different solutions can arise in the process of designing samples of complex equipment.

In this way, the analysis splits the overall function F of the created object into a number of simple subfunctions.

$$F \rightarrow \{F_{ijk}\}, \tag{5}$$

where i denotes the level of detail, j is the number of subfunctions in the level, and k is the number of subfunctions in the group.

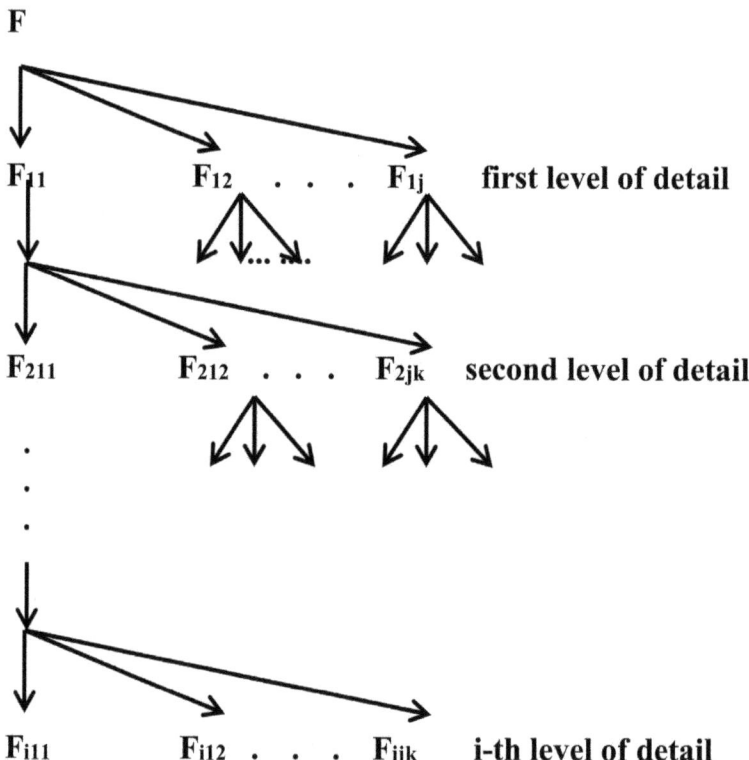

Figure 4. Analysis tree.

The analysis results in a set of elementary subfunctions that are not subject to further detail.

The elementarity property can be achieved for different subfunctions at different levels of detail. To determine the objective properties of an elementary subfunction (completion of the itemization process), the following rules can be used:
- A subfunction is elementary if its subsequent algorithmization is not difficult for the designer;
- A subfunction is elementary if the designer knows a variant of its in situ implementation that can be used as a readymade part of the object being created.

Due to the subjectivity of the analysis procedure, the result (a set of elementary subfunctions) is highly dependent on the skills and training of the designer, and it is highly questionable in terms of its optimality.

During the analysis of a generalized function, the designer must constantly monitor the correctness of the detailing, which consists of checking the implementation of the analyzed function by means of the set of subfunctions resulting from its detailing. For this purpose, it is sufficient to represent each subfunction in the form of Equation (4), and then analyze the set of subfunctions for joint implementation. In [11], we proposed a procedure of such analysis, which consists of "extinguishing" the internal operands and comparing the obtained result with the description of the analyzed function in the form of Equation (4). The absence of necessary operands or appearance of new ones, which are not used in the description of the detailed function, indicates incorrect specification or an error in assigning operands to subfunctions.

The efficiency of constructing an F-image of the projected object can be improved by using readymade variants of detail at the upper levels during the analysis. Indeed, experience shows that several initial levels of detail can be the same for similar objects. For example, in the design of most software systems, a generalized function at the first level of detail is represented as three invariant components:

1. Data entry;
2. Obtaining a result;
3. Output of the result.

This fact is reflected in HIPO diagrams [12], popular among software developers in the early stages of software design technology.

The generalization of experience in designing various objects will make it possible to identify invariant levels of F-image description for similar objects. This will significantly improve the quality of design, as the invariant levels define the design at the initial stage, where miscalculations are most likely to have serious consequences for the project as a whole.

In the course of designing a new object, the designer may encounter functions that are being implemented for the first time in engineering practice. For such functions, there may not be a corresponding physical principle of operation (PFA). To solve such problems in CAD, there should be special means of checking the existence of the physical principle of action and, if necessary, the synthesis of a new physical principle of action on the basis of known physical and technical effects (PTE) [13]. In the presence of the function description in the form of Equation (4), as well as a library of descriptions of existing PTEs, the task can be solved in an automated mode in accordance with the scheme in Figure 5.

As can be seen from the block diagram in Figure 5, the required PFA is synthesized as a "chain" of PTEs, which are described in some library. The "chains" are formed on the basis of correspondence of O-operands and I-operands of the considered PTEs. Several competing "chains" can be obtained as a result of solving the problem. As a criterion for choosing the best solution variant, one may use, for example, time, cost of implementation of a new PTE, and accuracy of the result obtained when using it in the course of design.

All newly synthesized FPPs are accumulated in the type library and can be used in the future.

To automate the task of synthesizing a new PFA, off-the-shelf tools can be used, examples of which are software packages such as Relko (Hilden, Germany), "Inventing machine" (Minsk, Belarus), Tech Optimizer (Oregon City, OR, USA), Product Function Definition (Washington, DC, USA), and Product Function Optimizer (Washington, DC, USA).

Concluding the overview of the procedure of building the F-image of the designed object, it should be noted that the analysis procedure is insufficiently formalizable; in modern CAD, it is performed in the "manual" mode. It can be improved by developing information support tools for the designer. In particular, it is advisable to use automated reference information about the results of designing similar or similar objects.

Procedures for controlling the correctness of the results of analysis and synthesis of new FTPs, although not present in modern CAD, are sufficiently well formalized and can be implemented in new designs.

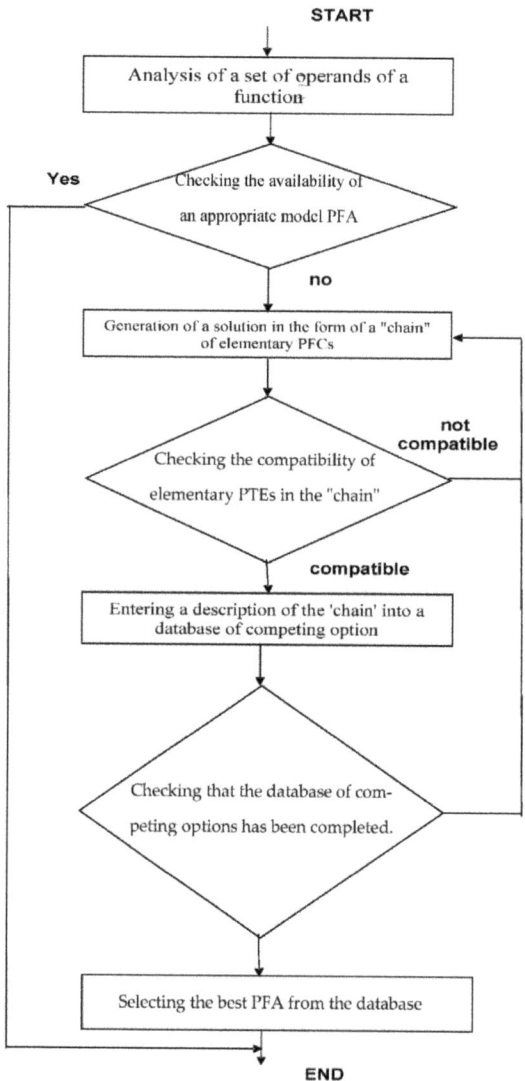

Figure 5. Synthesis scheme.

3. Results

One of the main provisions of the aircraft design methodology is to single out two levels of project development—external and internal design [13]. The fundamental difference between them is due to the difference in the final goals of the product development processes implemented here. The purpose of external design is to determine the feasibility and feasibility of creating a product, and the purpose of internal design is to obtain information necessary and sufficient to create a product under specified conditions. The internal design process begins with the overall concept of the aircraft and the shaping of its appearance. It is the task of shaping the appearance that serves as a connecting link in organizing the interaction of external and internal design systems, as shown in Figure 6.

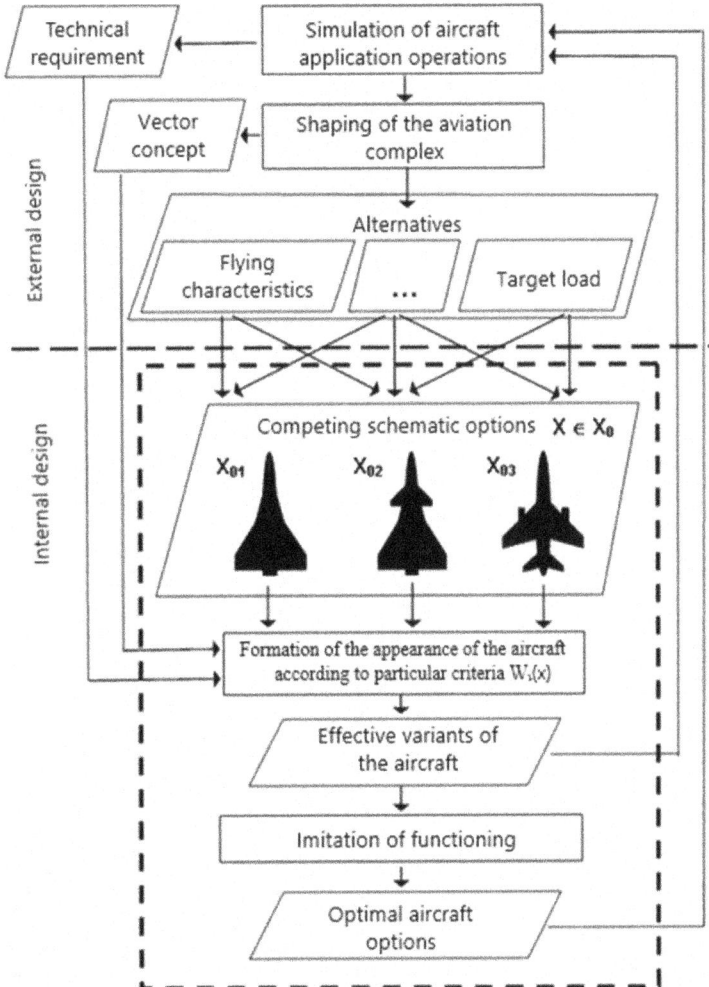

Figure 6. Model of interaction of external and internal design systems.

3.1. Aircraft Shaping: Traditional Methodology

The concept of "product appearance" does not have a strict definition, but most design specialists include structural (schematic) features and the most important parameters of the design object in its composition, which uniquely determine its shape, size, and takeoff weight. Schematic features include the aerodynamic scheme ("normal", "duck", or "tailless"), the location of the wing (lower, middle, or upper), the shape of the wing in plan, types of wing mechanization devices along the leading and trailing edges, the scheme plumage (low-lying or T-shaped), and type and location of engines. This group of features is called shape parameters. They define a "dimensionless" prototype of the aircraft, the dimensions of which are further determined by subsequent calculations. Within the framework of the selected combination of schematic features, the dimensions of the aircraft are determined primarily by the wing area and the starting thrust of the engines (or the specific load on the wing and the starting thrust-to-weight ratio derived from them). This group of parameters is referred to as dimension parameters.

The perfection of the chosen scheme is characterized by functions of the airframe geometry such as the lift coefficient $Cy\alpha$, drag coefficient $Cx\alpha$, aerodynamic quality K, and the efficiency of the power unit, which are functions of the gas dynamic characteristics of the engine, e.g., the specific fuel consumption Cp and the relative fuel mass mt [14].

Together, these parameters uniquely determine the flight characteristics of the aircraft. By varying them, the designer achieves the design goals.

Thus, the problem of shaping the appearance (AS) includes the subtasks of structural synthesis (determination of circuit solutions) and parametric synthesis (determination of the optimal values of the dimension parameters).

In this case, the problem of structural synthesis cannot be finally solved in isolation, since the efficiency of the obtained circuit solution can be confirmed only as a result of solving the problem of parametric synthesis.

It is this circumstance that makes the AS task so important and so difficult to formalize. The complexity of this task is not so much in organizing the enumeration of permissible combinations of circuit features, but in the difficulties of comparing synthesized variants of circuits.

The complexity of the considered problems is highlighted in Figure 6 with a dot-and-dash frame. At the same time, this work does not consider the formation of a geometric model of an aircraft; it can be built interactively or generated automatically in a specialized knowledge-oriented CAD subsystem, which does not affect its subsequent use in any way.

The traditional solution to the problem of selecting rational options for design solutions at the preliminary design stage is to use empirical or approximate analytical dependencies to determine the aerodynamic and flight characteristics [14]. The same path was proposed in the works of recent years [15,16]. A similar approach to the AS problem (search for analogs, enumeration of combinations of circuit features, and calculation of characteristics based on simplified dependencies) also prevailed in [17,18].

The motive for refusal from numerical research at the stage under consideration is usually its high labor intensity. For all the seeming naturalness of this approach (many initial data for accurate analysis have simply not been established yet), it has a serious drawback—the low accuracy of estimates of the aircraft's functional properties, i.e., its aerodynamic and flight characteristics.

At the same time, only flight characteristics can act as private criteria for evaluating and comparing options at the preliminary design stage. The main source of regulatory data is the airworthiness standards for aircraft of various categories (e.g., [19]), and practically the only category of aircraft properties that can be checked for compliance with the standards at this stage is flight characteristics, since these properties have the most complete quantitative representation in them. Therefore, it is quite natural to strive to obtain a more accurate assessment of these characteristics at the earliest possible stages of development. However, the well-known simplified dependencies cannot cover the entire wide range of speeds specified in the norms.

The feasibility of such a solution is supported by the leading trends in the development of software engineering analysis (CAE) in recent years, primarily the following:

- The transition to multidisciplinary (so-called multiphysics) modeling for solving related problems requiring the simultaneous analysis of processes of different physical nature, e.g., the gas flow around a structure and its deformation under the action of this flow; such capabilities can be both embodied in one multidisciplinary system, for example, VHDL-AMS simulators, SimulationX, ANSYS multiphysics, and NASTRAN multidisciplinary, or implemented through the interaction of multidisciplinary systems, such as FlowVision and ABAQUS;
- Transition from design verification to up-front simulation; the experience of using the SimDesigner 2004 toolkit in the CATIA V5 environment showed that transferring the analysis of design solutions to earlier design stages significantly increases the efficiency of CAD; the development of this direction was the SIMULIA project by Dassault Systemes, which laid the foundation for a new class of CAD tools—realistic

simulation, also called Rapid Analysis and Validation of Design Alternatives (RAVDA); similar developments were carried out by other companies;
- The use of CAE systems not only for verification, but also for the synthesis of design solutions; typical examples are the methods of optimization of the shape of objects (topology optimization) by iterative execution of procedures for the analysis of the stress state and subsequent exclusion from the model of the least loaded finite elements, as a result of which a structure close to equal strength is formed.

An additional incentive to include software tools for the analysis of aerodynamic processes, computational fluid dynamics (CFD), and flight simulation systems together with specialized programs for synthesizing the layout and geometry of the model in a single cycle of forming the appearance of the aircraft is another modern trend in the construction of CAD—the transition from "initially integrated" complexes to "freely integrated" sets of functional modules. This approach (arrangement of readymade modules with a minimum of own programming) is successfully used in the design of microelectronic and micromechanical products, where such systems are called heterogeneous CAD systems. This path seems to be promising for other areas of technology, which is confirmed by the developments of foreign research groups, particularly the work of the Delft Technical University [20].

Thus, the analysis of previous works indicates the existence of a contradiction between the need for multivariate design, the requirements for reducing the development time, and the insufficient accuracy of methods for assessing the aerodynamic and flight characteristics of an aircraft during preliminary design. To resolve this, it is proposed to use CFD systems and flight simulators at the earliest possible design stages from the moment of synthesis of the first variants of the topology and geometry of the aircraft, i.e., already in the problem of shaping the appearance.

The general sequence of design procedures in the proposed preliminary design methodology as a whole also corresponds to Figure 6; however, their content changes as follows:
- Selection of a combination of schematic features for the next design option:
- Design calculations of the main parameters of the aircraft;
- Building a geometric model of the first approximation;
- Virtual blowdown of the model in the CFD system;
- Determination of the main aerodynamic characteristics;
- Setting the initial data and/or programming the flight dynamics block of the flight simulator;
- Virtual test flights in the simulator;
- Conclusion about the possibility of achieving the specified characteristics and about compliance with airworthiness standards.

To test the proposed preliminary design technology, a prototype of the software package was developed as part of the SolidWorks CAD system, the Plane3D proprietary application for the automated synthesis of the geometric model, the Flow Vision CFD system, and the Flight Gear flight simulator. The complex also uses general-purpose software—the Notepad text editor for storing the coordinates of the points of standard aerodynamic profiles, the MS Excel spreadsheet processor for storing the characteristics of analog aircraft and calculation results, and the Blender graphic editor for converting the 3D model into the format required by the flight simulator. The scheme of the complex is shown in Figure 7.

Below, the main stages of work on the analysis of the flight characteristics of the designed aircraft are illustrated.

The geometric model of the aircraft, generated by the Plane3D application in cooperation with SolidWorks (Figure 8), is translated and transmitted to the Flow Vision CFD system in VRML format.

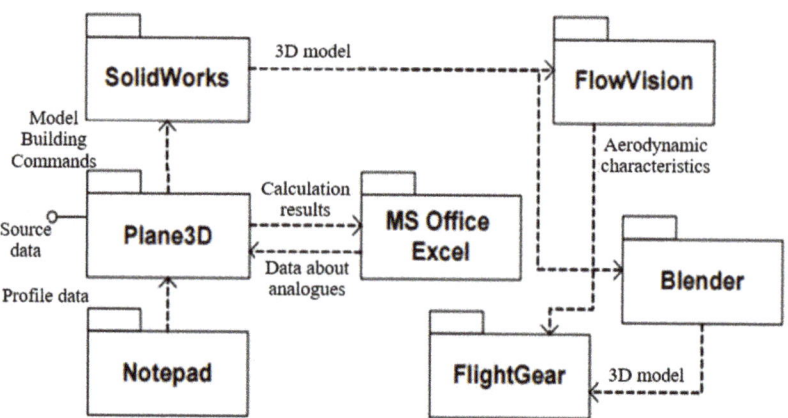

Figure 7. Block diagram of the software package.

Figure 8. Aircraft geometric model.

In the Flow Vision system, a mesh of finite volumes is generated (Figure 9), the conditions for adapting the mesh and rhe boundary and initial conditions are set, and then a virtual blowdown of the model is performed in the mode of interest to the researcher.

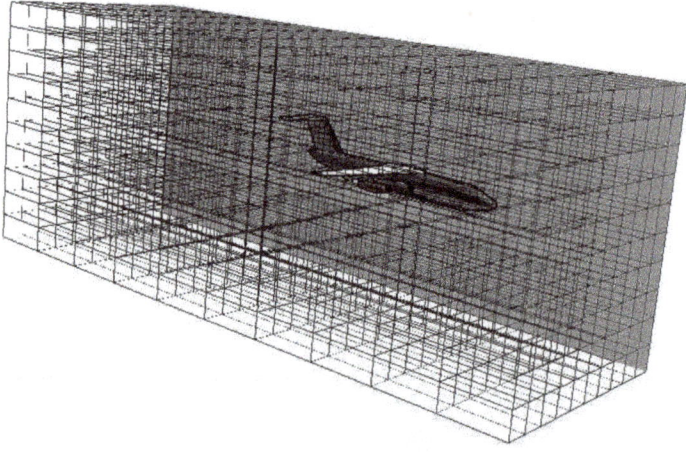

Figure 9. Final volume mesh in the Flow Vision system.

The results obtained—values or graphs of changes in flow rates, pressures, aerodynamic forces, and other parameters (Figures 10 and 11)—are subject to processing for the appropriate adjustment of the simulator dynamics block.

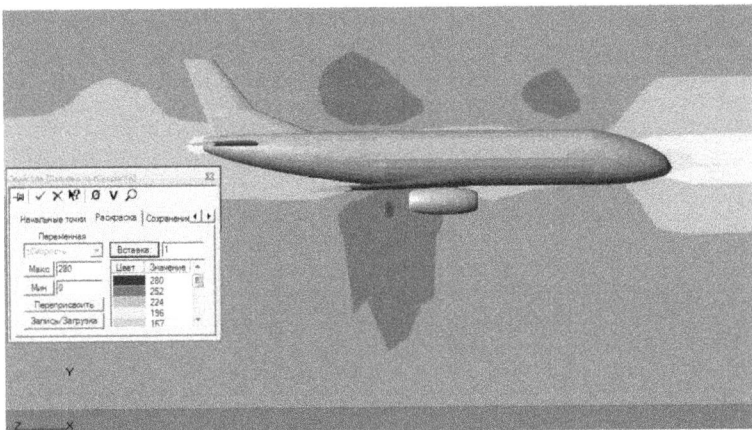

Figure 10. Velocity distribution in the plane of symmetry of the aircraft.

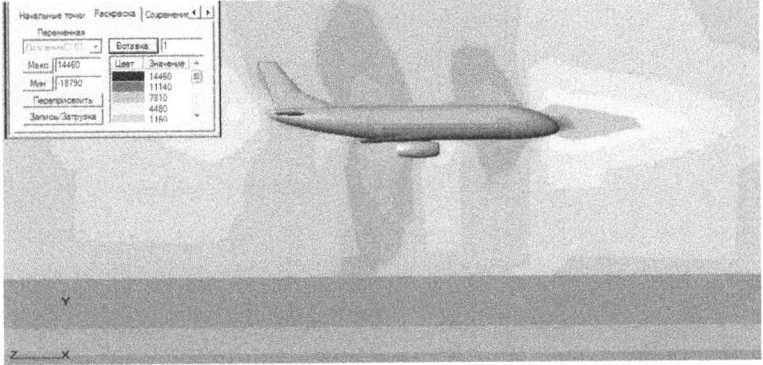

Figure 11. Pressure distribution in the plane of symmetry of the aircraft.

After completing the simulator setup, the model is converted to AC3D format and loaded into the simulator to perform virtual flight tests (Figure 12).

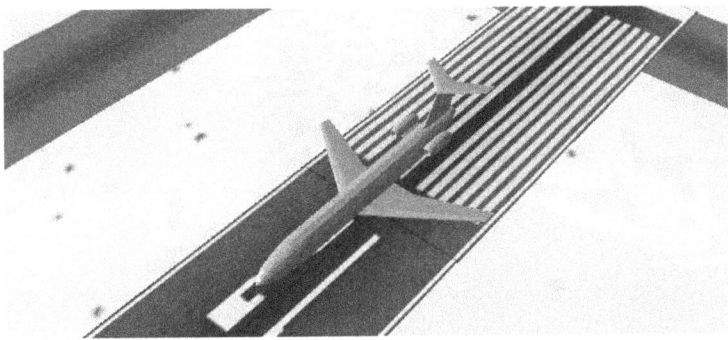

Figure 12. Airplane model loaded into the simulator.

3.2. Architecture of a Multiagent Platform for Structural–Parametric Synthesis of Objects

The design of complex technical objects requires the simultaneous consideration of a large number of relationships of various types (cause-and-effect, temporal, and spatial) between their elements and properties and processes of various physical nature (mechanical, electrical, hydraulic, etc.). The construction of a unified informational or mathematical model of such objects is practically impossible, since a different mathematical apparatus is required to describe various types of relations and connections—various methods of continual and discrete mathematics. The decomposition principle leads to the formation, instead of a single model, of a certain set of submodels (private models), each of which reflects a certain aspect of consideration, i.e., a particular point of view on the design object. The number and types of these sub-models can change when moving between hierarchical levels (product, assembly, assembly, and part) and development stages. For an aircraft, the most important of these submodels are geometric, weight, aerodynamic models, models of flight dynamics, power unit, layout and alignment, efficiency, and economic models [14]. Note that the listed models relate only to the functional aspect; in the tasks of resource (strength), structural, and technological design, the corresponding submodels are added to the general list.

However, dividing the description of a complex object into particular models and corresponding groups of tasks significantly simplifies the modeling process within each aspect, while significantly complicating the procedures for corroborating particular design solutions obtained within this framework. Decomposition of a technical system reduces the explicit complexity, but increases the so-called implicit complexity associated with the difficulties in determining the expected properties of the system by the characteristics of its elements, which is a manifestation of the emergence property of complex systems [14].

The entire set of tasks to be solved can be classified according to such criteria as the hierarchical level, aspect, and type of task (see Figure 13).

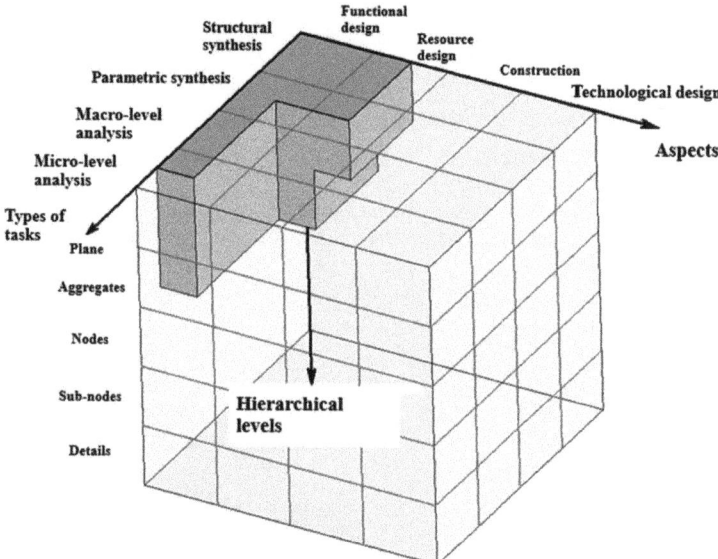

Figure 13. Classification of design tasks.

The figure shows an area that roughly corresponds to the tasks of forming the appearance of the product. Table 1 presents a list of design tasks for one of the units (the wing) as an example.

Table 1. Aircraft design tasks at the level of the "wing" unit.

Aspect	Structural Synthesis	Parametric Synthesis	Macro Level Analysis	Micro Level Analysis
Functional design	Determination of the wing shape, as well as the types and location of mechanization equipment	Calculation of the geometry of the wing and its parts (area and dimensions)	Calculation of aerodynamic characteristics (Cx, Cy, etc.)	Numerical simulation of the flow around a wing
Resource design	Determination of the structural scheme of the wing (types and number of load-bearing elements)	Design calculation of the dimensions of the load-bearing elements	Kinematic and dynamic analysis of moving elements	Numerical simulation of the stress–strain state of the wing elements
Construction	Layout of structural elements and equipment (fuel, mechanization drives, chassis, etc.)	Distribution of layout elements by cross-sections	Alignment calculation	
Technological design	Choice of the scheme of technological division and assembly	Calculation of assembly dimensional chains	Design for assembly	
	The choice of a sheet stamping scheme (drawing, covering, etc.), the number of transitions, etc.	Calculation of the dimensions of the workpiece, drawing forces, contact pressure, etc.	Verification calculations of forces, elongation ratio, accuracy, etc.	Numerical modeling of the process of shaping and stress-strain state of the workpiece

Along each axis of the "system cube" (hierarchical levels, types of tasks, and aspects), movement in only one direction is permissible, but the sequence of individual steps along different axes is not limited by anything other than the availability of the necessary initial data, and nowhere is it stipulated in which direction to perform the first steps, under what conditions to change the direction of movement, and how long it is generally permissible to move in one direction. In addition, we will encounter "linked" (connected) problems, the sets of variables of which intersect, as seen in aeroelasticity problems.

This means that the design strategy may not be as rigidly regulated as with a sequential "aspect" pass. However, this requires a different organization of the information and procedural components of the CAD software. In particular, a flexible management of the sequence of design procedures, close to an adaptive design strategy, can be provided by a multidimensional model of the design object. One of the modern trends in the development of CAD is exactly the desire of developers to build systems of interconnected models that characterize various aspects of describing the design object. An example of such an approach is [14], where the modules of weight design, aerodynamic, and strength calculations were combined. This area also includes the work of a number of foreign research groups, particularly the Delft Technical University (The Netherlands), where the knowledge-oriented multi-model generator (MMG) system was created, the diagram of which is shown in Figure 14. The system uses the facilities of the GDL (General-Purpose Declarative Language). The combined information model of the object covers aerodynamic, strength, and production and economic "layers".

To date, a number of theories have been developed related to solving the problem of multidimensional modeling, the origins of which can be seen in the general design theory of Yoshikawa–Tomiyama, and further traced in numerous modifications of the FBS theory (function—mode of operation—structure) [21]. These theories imply the interaction and origin of one aspect from another, which is absolutely correct in terms of the sequence of design phases. However, for each aspect, the concept of its own knowledge model is introduced and, as a consequence, transitions from one model to another are necessary at each transition to the next design phase.

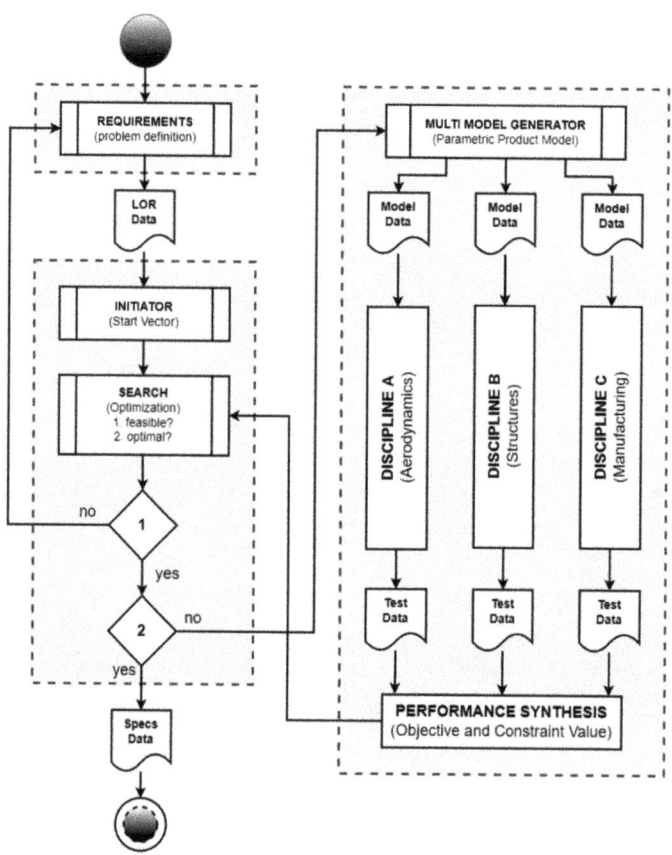

Figure 14. Multi-model generator knowledge-oriented system diagram.

For this reason, most theories are considered insufficiently formalized, since the list of knowledge representation models is rigidly defined and cannot be supplemented or changed. In fact, these theories are too formalized, since they formally describe each (but not all) phase and model in advance, which does not allow abstracting from specific design problems. Models and requirements should not be categorized as functional or structural.

If we define a certain format for describing requirements, which can be formalized not within the framework of the specifics of its description, but universally for all possible options for presenting and describing requirements, then the division of design into phases will be extremely indicative, while the work of the system will be carried out outside the design phases.

It can be assumed that the main reason for dividing the design process into phases and stages is, apparently, a person's defensive reaction to two main factors: (a) insufficient amount of human memory, which requires the participation of different specialists to work on one rather complex project; (b) the need to agree on private ("one-aspect") decisions made by different developers outside of contact with each other.

Thus, transferring to the area of computer-aided design all the traditions of organizing the design of the non-automated, a person imposes on the computer their imperfect method of work, generated by the limitations not of the computer, but of its capabilities.

In the design methods of most technical objects, the sequence of stages is set rather rigidly, e.g., for an aircraft: aerodynamics → flight dynamics → structure and strength → technology . . . ; for an electric motor: electromagnetic calculation → thermal calculation → ventilation calculation → noise and vibration calculation → mechanical calculation → reliability calculation, etc.

For an automated system, both of the above factors are not decisive (provided that the procedures for agreeing decisions are also formalized); it really could work "outside the design phases" and there is no need for it to establish the sequence of determining the properties of various groups so strictly. However, this requires completely different design methods that do not imply the division of properties into categories (functional, constructive, costly, etc.) or by physical nature (mechanics, electrodynamics, hydraulics, etc.). It is still difficult to assess the harm from unnecessary restrictions, but it is obvious that everything that constrains the designer's freedom, whether in the content of actions or in their sequence, is not good for the cause.

Thus, it is possible to develop the most flexible system in which the work with different kinds of requirements will be uniform. By delegating calculations and checking compliance with the requirements for specific agents, we get a "design constructor" that any subject matter specialist can use, while accumulating their knowledge in the knowledge base.

Each element of this knowledge base is an agent, which has its own methods of calculation and verification of compliance with requirements. Therefore, when redesigning, the work of the designer is greatly simplified due to the reuse of previously designed elements. At the same time, the properties of an element are not divided into structural and functional. Properties are just parameters of an element, the main function of which is to determine compliance with the requirements; they are only secondarily responsible for visualizing the result—a physical or functional model of the designed product.

From this point of view, various types of design (structural, functional, and conceptual) cease to be decisive factors in choosing a design strategy, including computer-aided design. Multidimensionality cannot be achieved as long as the number of aspects is determined by the programmer developing software for domain specialists. It is necessary not to expand the number of approaches (aspects) to design (this tendency can be seen in a number of works, e.g., the general design theory of Yoshikawa–Tomiyama, i.e., design according to the requirements of the structure, and then the expansion of this theory by the functional aspect of Braha, i.e., paired design [22]), but to develop one universal approach that will define N aspects, where N depends directly on the end user.

The problem of moving from one stage of design to another is an illusion introduced into modern design by the need to recreate the image of the designed product from every point of view, forming a new model for each facet of the design solution. A technologist and a designer, working on one task, see it differently, but the essence of the task does not change from this; if the knowledge of the technologist and the designer were combined, then the design of two models with subsequent coupling and all the resulting difficulties would be meaningless.

Note that, from this point of view, the stage of ensuring the manufacturability of a product design (MPD) between design and technological preparation of production, legalized by many standards, is one of the most odious results of our artificial transition to sequential design; its main content is precisely the revision of part of the design solutions, accepted without taking into account the opinions of technologists (i.e., within one aspect). Here, one can see the loudly condemned "redesign" by all, but raised all over the world to the rank of a mandatory procedure.

The advantage of an automated system is that it can contain an unlimited amount of knowledge in various subject areas, while operating with one model. In this case, it is not necessary, as it is stated in Braha's works, to first check the compliance with the structural requirements, and then look for intersections with the satisfaction of functional requirements; one only needs to ensure that all requirements are met. Moreover, when solving a particular problem, there will still be a need for additional aspects.

By and large, functional design is just the definition of a list of functional properties (requirements) that the designed object must meet [23]. Even if it is possible to find a solution to the functional puzzle that will fully meet the requirements, there is no guarantee that the structure corresponding to the selected functional solutions will satisfy the structural requirements. Therefore, it is necessary to consider structural, functional, and other requirements as a single entity.

Consequently, the design system should consist of agents with a number of properties, the division of which into any categories is very arbitrary and introduced solely for the convenience of the end user. Each agent has calculation methods and methods for checking compliance with a certain requirement. In addition, the system has a number of requirements; each agent may or may not meet any requirement. However, parametric and structural constraints are also requirements. In such a system, design is reduced to the selection of a set of agents that meet all the requirements. The task of calculating the parameters and structure is a nested selection task, since, when determining the fulfillment or nonfulfillment of the requirement, all possible implementations of the agent are calculated. The sequence of design phases is implicitly taken into account when calculating the model, since previously defined properties are usually used to determine the value of a property. In general, the properties of aspects of earlier design phases are initially determined, e.g., from function to structure.

In accordance with our requirements, an agent is a highly mobile entity that can stop its execution, change its internal structure, and continue execution. The enlarged physical structure of an agent is shown in Figure 15.

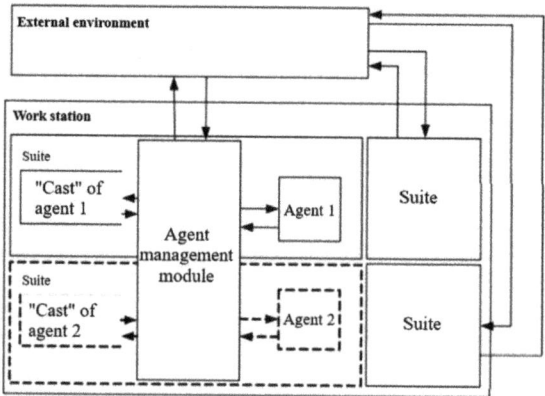

Figure 15. Enlarged agent structure.

The definitions of each entity in the figure are provided below.

Let us define an agent as a kind of autonomous entity that is capable of sensing and acting through receptors and effectors. In this case, the software implementation, perception, and impact will be carried out through the transmission of messages. When defining an agent, many people focus on intellectuality. The property of intellectuality depends on the specific implementation of the agent's behavior and does not directly depend on its architecture. The architecture in this case should be at least such that it allows one to show intellectuality. In this case, this can be achieved through input/output channels (sensation/impact).

The external environment is defined as a set of objects that do not belong to a given agent and are perceived as a separate entity. Signals are received from there, which the agent can ignore if it does not know the kind of signals or respond with some action. Thus, if the external environment is defined as a set of agents, then, for an agent without receptors and effectors, this set will be empty. A larger set indicates that the agent knows more about the environment.

In this case, any agent must have the basic functionality defined in the agent management module; hence, each agent knows about the existence of other agents in its reach. The basic functionality of the agent with the IUnknown interface of the COM architecture can be compared, and then some analogy between the methods defined by this interface can be drawn.

The agent management module is a static part of the agent and is responsible for the implementation of the agent specification in memory, transferring incoming messages and events to it, as well as sending messages to the external environment. It is necessary to stipulate right away that synchronization with other agents, blocking access to the agent, resolving agent links, stopping and starting the agent, and saving its state will take place at this level.

The "cast" of the agent involves the designation of all metadata and dynamic data about an object, i.e., its specification, the last saved state, and associated data.

The agent's suite is the set of data and working modules that relate to a specific agent.

It is necessary to clarify the essence of the relationship between the agent management module and the agent itself. Figure 15 shows the case when one control module can manage several agents. Here, one can draw an analogy with a CORBA server, which, when receiving a new request, sends it to the addressee—a specific instance of an object. In this case, the situation is somewhat different, but the essence remains unchanged; in fact, the agent management module is its server part, which sends messages intended for it to the agent's domain. The creation of a new agent can occur as with cloning the mobile part (e.g., if it is necessary to move a new agent to another runtime, roughly speaking, to another computer) or without cloning the mobile part, where several agents will correspond to one control module. However, below, we take the control module and the agent as a whole. The cloning process of a control module must be transparent to the ultimate creator of the agent. Adopting such an architecture, we get a mobile agent, for which no additional components are needed.

Obviously, even the most experienced specialist cannot handle the representation of integrity when creating complex systems. This is where the myth of the inevitability of dividing the design process into aspects with different models originates. However, an automated system can have a knowledge base of unlimited complexity, and the solution to the problem of choosing an engine for an aircraft will be carried out according to the same algorithm as the choice of a simple mechanism.

The choice of the type of engine can be made on the basis of the dependence of the thrust efficiency on the number M of the flight (Figure 16).

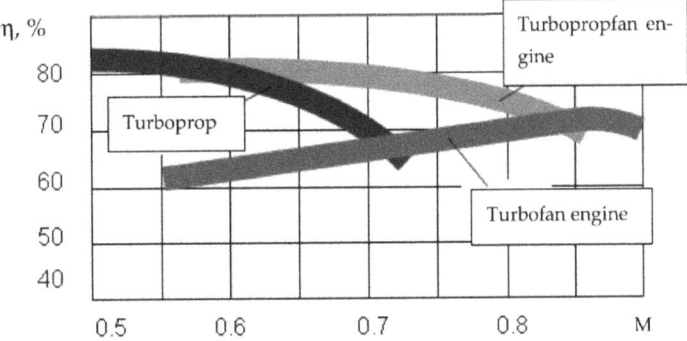

Figure 16. Dependence of the thrust efficiency of the engine on the number.

The mechanism of the corresponding module of the knowledge base is the decision table (Table 2). At the same time, the choice of an agent should not be understood as the choice of a finished result from an already existing list.

Table 2. Decision table for determining the type and characteristics of the engine.

Flight Number M	Up to 0.55	0.55–0.83	Over 0.83
Engine type	Turboprop	Turboprop fan engine	Turbofan engine
...	...	...	...
Diameter of the engine (propeller), m	$(N_0/8.1)^{0.25}$	...	$0{,}1 \cdot G_B^{1.5} \cdot \varphi_1(R_0) \cdot \varphi_2(\pi_K^*) \cdot \varphi_3(T_3)$

Note: N_0—power; G_B—air consumption in a second; R_0—starting thrust; π_K^*—the degree of pressure increase in the compressor; T_3—the gas temperature in front of the turbine; φ_i—statistical coefficients.

The calculation of the parameters and the determination of the structure in the design is also a kind of choice of values or elements of the product, such as, for example, the calculation of the dimensions of the engine in Table 2. In the case of the multidimensional agent network (knowledge base) described above, the choice of the calculation method is added to these selection parameters, i.e., the choice of a solution in the form of one or another agent.

For an automated system, the traditional division of tactical and technical requirements (TTT) into tactical and technical ones is irrelevant; not only technical requirements, but also a number of constructive ("structural") features are extracted from tactical (in the general engineering sense of "functional") requirements in one iteration, e.g., using production rules of the following form:

IF TTTs provide for the standard conditions of use of the aircraft (for example, subsonic passenger or transport), THEN the structure of the aircraft should include fuselage, wing, empennage, and landing gear.

IF the given cruising speed significantly exceeds the landing speed (determined by the class of the aerodrome specified in the TTT), THEN the wing structure should include the flap together with its attachment units.

Hence, it follows that in the conditions of computer support for making design decisions, there is no need to delay the start of the work of designers until the complete appearance of the product, whereas, for technologists, there is no need to delay the start of the work until the preparation of working documentation. The task of modern developers is to overcome the barrier of formalization of knowledge, to present it as a certain entity corresponding to a single, flexible, and scalable pattern.

4. Discussion

The considered approach was experimentally used in the development of a small aircraft at the Institute of Physical Modeling Problems of the National Aerospace University "Kharkov Aviation Institute". Table 3 shows the planned indicators in the preparation of the project.

It was revealed that, during the implementation of the project, temporary changes should be made to the organizational structure to reduce the time for coordinating decisions, thereby reducing permits. It is necessary to make the following design changes in the organizational structure of the enterprise for the duration of the project: provide an economist to the project team to transfer the sector of strength, aerodynamics, power plant, and technology to the subordination of the design team, to the subordination of the chief designer the department of avionics and control systems, and to an experimental team, as well as create conditions for the possibility of using pilot production through the chief designer. This does not mean that it is necessary to disband the existing divisions, but an internal order should be created, according to which, for the duration of the project, there will be such an organizational structure of the project itself. The actual data obtained during the implementation of the project are presented in Table 4.

Table 3. Planned indicators of the project.

Stage Name	Stage Duration, Months	Duration in Relation to the Entire Project %
Marketing research	1	3.33
Rationale for the possibility of creating	2	10.00
Avan project (feasibility study)	2	16.67
Submission of an application to the state aviation agency	1	20.00
Preliminary design	2	26.67
Creation of a layout	0.25	27.50
Creation of a technical product	2	34.17
Creation of working design documentation	2	40.83
Preparing for the production of a prototype	1	44.17
Prototype production	6	64.17
Test preparation	1	67.50
Flight development tests	6	87.50
Correction of design documentation	0.25	88.33
Certification tests	2	95.00
Documentation approval	1	98.33
Obtaining a type certificate	0.5	100

Table 4. Factual data.

Stage Name	Stage Duration, Months	The Duration of the Project in Relation to the Planned Dates %
Marketing research	1.00	3.33
Rationale for the possibility of creating	2.00	10.00
Preliminary project (feasibility study)	2.00	16.67
Submission of an application to the state aviation agency	1.00	20.00
Preliminary design	5.50	38.33
Creation of a layout	0.25	39.17
Creation of a technical product	1.00	42.50
Creation of working design documentation	1.00	45.83
Preparing for the production of a prototype	1.00	49.17
Prototype production	4.50	64.17
Test preparation	1.00	67.50
Flight development tests	1.25	71.67
Correction of design documentation	0.00	71.67
Certification tests	0.75	74.17
Documentation approval	1.00	77.50
Obtaining a type certificate	0.50	79.17

It should be noted that products of complex technology are, first of all, unique; therefore, projects for their creation are quite unique in nature.

From the data obtained, it can be seen that the actual data throughout the lifecycle receive a significant acceleration at those stages that are associated with the perception of the initial data and the performance of work without the accompanying errors associated with a misunderstanding of the product requirements. Specialists choose the description they need and perform the tasks assigned to them more efficiently. However, such good results can also be associated with the experience of implementing similar projects among specialists from each department. For each design bureau, the percentage of time saved will be different, ranging from 7% to 21%.

5. Conclusions

The design procedure is invariant for a wide class of engineering objects and represents the construction of a complete description of the designed object, consisting of three components: F-description, M-description, and N-description.

The leading component of the description is the F-description of the object.

The procedure for constructing a functional description is formalizable and can be performed in automated mode. It is advisable to use a standardized procedure in the design of the software to support the LRC. During testing of the prototype of the complex, the main results were obtained, which are given below.

The possibility of including the problems of aerodynamics analysis and flight simulation in a single aircraft design process at the early stages of development was established.

When using the described approach, the development time was reduced to 7% if the development was completed by a preliminary project. In the case of using the proposed approach throughout the entire lifecycle of the project, a reduction of up to 21% was achieved.

The need for a more thorough construction of a geometric model than is usually accepted in the tasks of shaping the appearance is revealed due to the sensitivity of CFD systems to the smoothness of the model surfaces.

The main difficulties that prevent the complete automation of the task of forming the appearance of the aircraft are as follows:

- Lack of objective information about the accuracy of simulation results in various simulators; this may require separate studies to compare the capabilities of different flight simulation systems;
- Absence (for objective reasons) at the early stages of development of a number of initial data for the simulator (these data, in principle, cannot yet be obtained); this may require searching for a prototype by morphological methods (e.g., [18]) and using its characteristics as a first approximation;
- Labor intensity processing the results of aerodynamic analysis and preparing data for the simulator on the basis of the results of virtual blowdown; this may require the creation of translator programs;
- Lack of a unified format for presenting the aerodynamic and flight characteristics of an aircraft.

Author Contributions: Conceptualization, V.S. and D.K.; methodology, D.K.; software, O.P.; validation, V.S., D.K. and O.P.; formal analysis, V.S.; investigation, D.K.; resources, O.P.; data curation, O.P.; writing—original draft preparation, D.K.; writing—review and editing, V.S.; visualization, V.S.; supervision, D.K.; project administration, D.K.; funding acquisition, O.P. All authors have read and agreed to the published version of the manuscript.

Funding: This research received no external funding.

Institutional Review Board Statement: Not applicable.

Informed Consent Statement: Not applicable.

Data Availability Statement: Not applicable.

Conflicts of Interest: The authors declare no conflict of interest.

References

1. Ek, K.; Mathern, A.; Rempling, R.; Brinkhoff, P.; Karlsson, M.; Norin, M. Life Cycle Sustainability Performance Assessment Method for Comparison of Civil Engineering Works Design Concepts: Case Study of a Bridge. *Int. J. Environ. Res. Public Health* **2020**, *17*, 7909. [CrossRef] [PubMed]
2. Pinheiro Melo, S.; Barke, A.; Cerdas, F.; Thies, C.; Mennenga, M.; Spengler, T.S.; Herrmann, C. Sustainability Assessment and Engineering of Emerging Aircraft Technologies—Challenges, Methods and Tools. *Sustainability* **2020**, *12*, 5663. [CrossRef]
3. Müller, J.R.; Panarotto, M.; Isaksson, O. Design Space Exploration of a Jet Engine Component Using a Combined Object Model for Function and Geometry. *Aerospace* **2020**, *7*, 173. [CrossRef]
4. Kozma, D.; Varga, P.; Larrinaga, F. System of Systems Lifecycle Management—A New Concept Based on Process Engineering Methodologies. *Appl. Sci.* **2021**, *11*, 3386. [CrossRef]
5. Kawakami, K.; Fukushige, S.; Kobayashi, H. A functional approach to life cycle simulation for system of systems. *Procedia CIRP* **2017**, *61*, 110–115. [CrossRef]
6. Koval, V.V. Application of modularity for normative documentation design. *Technol. Audit. Prod. Reserves* **2013**, *5*, 25–27. [CrossRef]
7. Kritskiy, D.N.; Druzhinin, E.A.; Karatanov, A.V.; Kritskaya, O.S. Content Management Method of Complex Technical System Development Projects. In *Advances in Intelligent Systems and Computing IV. Advances in Intelligent Systems and Computing*; Shakhovska, N., Medykovskyy, M.O., Eds.; Springer: Cham, Switzerland, 2020; Volume 1080, pp. 293–303. [CrossRef]
8. Polovinkin, A.I. Fundamentals of Engineering Creativity. In *Textbook for Students of Higher Education Institutions*; Mechanical Engineering: Moscow, Russia, 1988; p. 368.
9. Hubka, V. *Theory of Technical Systems*, 2nd ed.; Springer: Berlin/Heidelberg, Germany, 1988; p. 278. [CrossRef]
10. Ylmén, P.; Berlin, J.; Mjörnell, K.; Arfvidsson, J. Managing Choice Uncertainties in Life-Cycle Assessment as a Decision-Support Tool for Building Design: A Case Study on Building Framework. *Sustainability* **2020**, *12*, 5130. [CrossRef]
11. Shevel, V.V. Formalized procedure of consistency control of conjugate program units. In *Theory of Computer-Aided Design*; KhAI Publishing House: Kharkiv, Ukraine, 1984.
12. Nazarov, S.V. *Architectures and Design of Software Systems: Monograph*, 2nd ed.; INFRA-M: Thane, India, 2016; p. 374. [CrossRef]
13. Frederic, B.; Christopher, F.; Mavris, D.N. An Aircraft Development Methodology Aligning Design and Strategy to Support Key Decision Making. AIAA 2016-1661. In Proceedings of the 57th AIAA/ASCE/AHS/ASC Structures, Structural Dynamics, and Materials Conference, San Diego, CA, USA, 4–8 January 2016; p. 1661. [CrossRef]
14. Kritskiy, D.; Yashin, S.; Koba, S. Unmanned Aerial Vehicle Mass Model Peculiarities. In *Mathematical Modeling and Simulation of Systems (MODS'2020). MODS 2020. Advances in Intelligent Systems and Computing*; Shkarlet, S., Morozov, A., Palagin, A., Eds.; Springer: Cham, Switzerland, 2021; Volume 1265, pp. 299–308. [CrossRef]
15. Gorbunov, A.A.; Pripadchev, A.D.; Elagin, V.V.; Magdin, A.G.; Ezerskaya, E.M. The Method of Selecting Aircraft Conceptual Design Parameters at the Stage of Feasibility Study. *Int. J. Eng. Trends Technol.* **2021**, *69*, 193–198. [CrossRef]
16. Saporito, M.; da Ronch, A.; Schmollgruber, P.; Bartoli, N. Framework development for robust design of novel aircraft concept. In Proceedings of the 3AF Aerospace Europe Conference, Bordeaux, France, 25 February 2020; pp. 1–11.
17. Rentema, D.; Jansen, E. An AI tool for conceptual design of complex products. In *Design Research in the Netherlands*; Eindhoven University of Technology: Eindhoven, The Netherlands, 2000; pp. 119–131.
18. Raymer, D.P. Enhancing Aircraft Conceptual Design Using Multidisciplinary Optimization. Ph.D. Thesis, Royal Institute of Technology, Stockholm, Sweden, 2002.
19. Aviation Rules of Ukraine. Available online: https://avia.gov.ua/discussion_category/aviatsijni-pravila-ukrayini/ (accessed on 20 June 2022).
20. Schut, J.; van Tooren, M.; Berends, J. Feasilization of a structural wing design problem. In Proceedings of the 49th AIAA/ASME/ASCE/AHS/ASC Structures, Structural Dynamics, and Materials Conference, 16th AIAA/ASME/AHS Adaptive Structures Conference, 10th AIAA Non-Deterministic Approaches Conference, 9th AIAA Gossamer Spacecraft Forum, 4th AIAA Multidisciplinary Design Optimization Specialists Conference, Honolulu, HI, USA, 5 August 2008; p. 2263.
21. Tomiyama, T.; Yoshikawa, H. *Extended General Design Theory*; Department of Computer Science; CWI: O'Connor, WA, Australia, 1986; p. 32.
22. Braha, D.; Reich, Y. Topological structures for modeling engineering design processes. *Res. Eng. Des.* **2003**, *14*, 185–199. [CrossRef]
23. Kritsky, D.N.; Druzhinin, E.A.; Pogudina, O.K.; Kritskaya, O.S. A Method for Assessing the Impact of Technical Risks on the Aerospace Product Development Projects. In *Advances in Intelligent Systems and Computing III. CSIT 2018. Advances in Intelligent Systems and Computing*; Springer: Cham, Switzerland, 2019; Volume 871, pp. 504–521. [CrossRef]

Article

A Linear Elasticity Theory to Analyze the Stress State of an Infinite Layer with a Cylindrical Cavity under Periodic Load

Vitaly Miroshnikov [1],*, Basheer Younis [1], Oleksandr Savin [1] and Vladimir Sobol [2]

[1] Department of Aircraft Strength, National Aerospace University "KHAI", Chkalova str., 17, 61000 Kharkiv, Ukraine
[2] Department of Theoretical Mechanics, Mechanical Engineering and Robotic Systems, National Aerospace University "KHAI", Chkalova str., 17, 61000 Kharkiv, Ukraine
* Correspondence: v.miroshnikov@khai.edu; Tel.: +380-67-7893333

Abstract: The design of parts of machines, mechanisms, structures and foundations, particularly in the aerospace industry, is closely related to the definition of the stress state of the body. The accuracy of determining the stress state is the key to optimizing the use of materials. Therefore, it is important to develop methods to achieve such goals. In this work, the second main spatial problem of the elasticity theory is solved for a layer with a longitudinal cylindrical cavity with periodic displacements given on the surface of the layer. The solution of the problem is based on the generalized Fourier method for a layer with a cylindrical cavity. To take into account periodic displacements, an additional problem is applied with the expansion of the solution for a layer (without a cavity) in the Fourier series. The general solution is the sum of these two solutions. The problem is reduced to an infinite system of linear algebraic equations, which is solved by the reduction method. As a result, the stress-strain state of the layer on the surface of the cavity and isthmuses from the cavity to the boundaries of the layer was obtained. The conducted numerical analysis has a high accuracy for fulfilling the boundary conditions and makes it possible to assert the physical regularity of the stress distribution, which indicates the reliability of the obtained results. The method can be applied to determine the stress-strain state of structures, whose calculation scheme is a layer with a cylindrical cavity and a given periodic displacement. Numerical results make it possible to predict the geometric parameters of the future structure.

Keywords: layer with a cylindrical cavity; Lamé equation; generalized Fourier method; Fourier series; reduction method

Citation: Miroshnikov, V.; Younis, B.; Savin, O.; Sobol, V. A Linear Elasticity Theory to Analyze the Stress State of an Infinite Layer with a Cylindrical Cavity under Periodic Load. *Computation* **2022**, *10*, 160. https://doi.org/10.3390/computation10090160

Academic Editor: Demos T. Tsahalis

Received: 5 August 2022
Accepted: 11 September 2022
Published: 14 September 2022

Publisher's Note: MDPI stays neutral with regard to jurisdictional claims in published maps and institutional affiliations.

Copyright: © 2022 by the authors. Licensee MDPI, Basel, Switzerland. This article is an open access article distributed under the terms and conditions of the Creative Commons Attribution (CC BY) license (https://creativecommons.org/licenses/by/4.0/).

1. Introduction

The main task in designing in the aerospace industry is the optimization of a unit and mechanisms. However, the geometry, constituent elements and type of load in most cases do not allow one to use accurate calculation methods. In this case, it is necessary to use time-consuming tests [1], in which the changing of geometrical characteristics leads to changes in physical and mechanical properties and new tests being required. The other way is to simplify the calculation model or to obtain the stress-strain state of the body by approximate methods [2]. The last two options do not provide certainty in the calculations results and are compensated by additional safety factors. Therefore, it is important to have a method that will allow one to obtain the stress-strain state of the body with a higher accuracy, taking into account the geometry and existing stress concentrators.

Approximate, exact and experimental methods are often combined with one another to increase the accuracy of the calculation. Thus, analytical, numerical and experimental research was carried out for a multilayer composite with a perpendicularly located cylindrical hole [3]. In an analytical study, the point stress criterion is developed. The finite element method is used in a numerical study. In [4], the analysis of the stress state for a plate with a cylindrical hole was carried out. The solution of the problem is based on metaheuristic optimization algorithms around stress concentrators. In paper [5], a plate with a cylindrical hole is also considered, but here the semi-analytical polynomial method is used. In this case, nonlinear partial differential equations were transformed into a system of nonlinear algebraic equations, and the Newton-Raphson method was applied. In [6], the complex potential method was used to study the bending of finite isotropic rectangular plates with a circular cutout. Combinations of described methods [3–6] effectively solve the problem with a perpendicularly located stress concentrator, but their application is not possible for longitudinal inhomogeneities and for cases of periodic loading.

In work [7], in order to calculate the reaction of a layered composite to an impact load, an analytical-experimental approach is proposed. In this case, analytical modeling is based on the decomposition into a power series of the component of the displacement vector in each layer for the transverse coordinate. In the experimental part, the maximum deflections of composite samples during the impact of the indenter are considered. In work [8], an analysis of the strength-laminated windows of an airplane cabin against a bird strike was carried out. The calculation method is based on embedding the original non-canonical shell into an auxiliary canonical form in a plane with boundary conditions. As a result, a simple analytical problem in the form of a trigonometric series is formed. An experimental model was developed to simulate the process of a bird hitting a hard target [9]. In work [10], the first-order theories for the analytical model of multilayer glazing are improved; they take into account the transverse shear deformations, thickness reduction and normal inertia of rotation of the elements of each layer. The mathematical model of the pressure pulse was experimentally investigated. The cited works [7–10] effectively determine the strength of layered composites but do not allow one to take longitudinal inhomogeneities into account.

To take into account longitudinal inhomogeneities, analytical or analytical-numerical methods use the Fourier series decomposition. Thus, in [11], steady-state problems of wave diffraction in a plate and a layer are solved on the basis of Fourier series. The paper [12] defines the diffraction of waves in space, half-space and an infinite layer that have a cavity or inclusion. The stress state for a layer with a cylindrical cavity or inclusion is considered in [13]. In [14], a layer with a cylindrical cavity or inclusion is considered, and the image method is used to solve the two-dimensional boundary value problem of the diffraction of symmetric normal longitudinal shear waves. These approaches [11–14] make it possible to obtain a highly accurate stress distribution or wave diffraction result for problems in a flat setting. However, these methods cannot be applied to a spatial problem and a problem with many boundary surfaces.

For a high-precision determination of the stress state of spatial models and for models with more than three boundary surfaces, the analytical-numerical generalized Fourier method [15] is most powerful. The paper [16] presents the substantiation of this method for the basic solutions of the Lamé equation for a half-space and a cylinder, written in the Cartesian and cylindrical coordinate systems, respectively. This makes it possible to solve the problem with flat and cylindrical surfaces.

Using the generalized Fourier method, a number of problems for a cylinder with cylindrical cavities or inclusions were solved. Thus, in [17], the problem was solved for a cylinder with four cylindrical cavities. The problem for a cylinder with N cylindrical cavities is considered in work [18], for a cylinder with cylindrical cavities forming a hexagonal structure in work [19], and for a cylinder with 16 cylindrical inclusions in work [20]. These papers [17–20] only apply the addition theorems of the generalized Fourier method for several cylindrical coordinate systems. This effectively solves problems for combinations of cylindrical surfaces but does not provide the ability to solve problems for a layer.

The use of the addition formulas of the generalized Fourier method for solutions of the Lamé equation between Cartesian and cylindrical coordinate systems is taken into account in other works. Thus, in [21], elasticity theory problems for a half-space with cylindrical cavities in displacements were solved. The mixed type is solved in [22]. In [23], the lower limit is taken into account, and the problem is solved for a layer with one cylindrical cavity in displacements. A problem with given stresses on the boundaries is solved in [24] and one of mixed type is solved in [25]. A layer conjugated with a half-space that has a cylindrical cavity is considered in [26]. A layer with one continuous inclusion is solved in paper [27] and one with two continuous inclusions is solved in paper [28]. In works [21–28], the double Fourier integral is applied to boundary conditions, which limits the range of use to rapidly decreasing functions only. This makes it impossible to take into account periodic displacements or stresses (for example, from located equipment or technological fastenings) when the loads are applied periodically through some constant interval of distance to infinity.

Therefore, the problem for a layer with a cylindrical cavity can be solved with high accuracy using the analytical-numerical generalized Fourier method. However, unlike existing works, it is necessary to apply additional methods to account for periodic loads and it is necessary to combine these methods.

Despite the large number of publications on the subject of calculating the layer with stress concentrators, there are in practice still many unresolved problems that occur during the design. The absence of a method for calculating such problems allows one to assert the relevance of conducting a study dedicated to the spatial calculation of a layer with a cylindrical cavity and specified periodic displacements. It should be noted that, in this paper, only linear elastic materials are considered.

2. Problem Statement, Materials and Methods of Analyzing the Stress State of the Layer

The object of research is the stress-strain state of a layer with a longitudinal circular infinite cylindrical cavity (Figure 1). The distance from the center of the cavity with a radius R to the upper boundary of the layer is $y = h$, and the distance to the lower boundary is $y = \tilde{h}$. The distances h and $\tilde{h}$ can have any value greater than R ($h > R < \tilde{h}$). A static load acts on the layer in the form of displacement (the second main task of the elasticity theory). In the theoretical formulation, the displacements at the upper boundary of the layer are taken into account in the form of periodic functions $U_x^{(h)}$, $U_y^{(h)}$, $U_z^{(h)}$ along the x-axis. The material of the layer is elastic, isotropic and homogeneous with linear elastic characteristics.

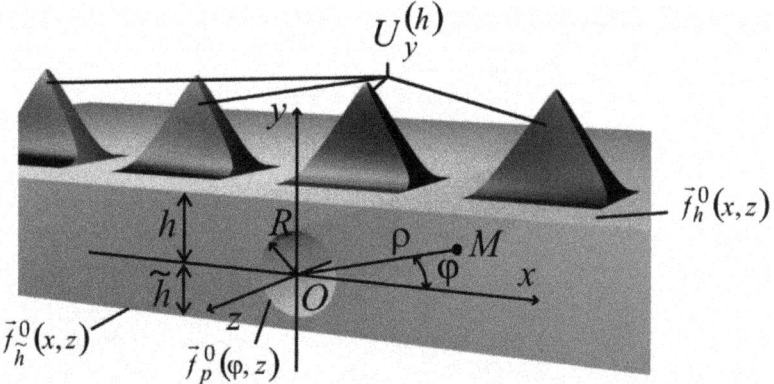

Figure 1. A layer with a cylindrical cavity and displacements specified on the boundary surfaces.

The cavity does not intersect with the boundaries of the layer and is considered in the cylindrical coordinate system (ρ, φ, z). The layer is considered in the Cartesian coordinate system (x, y, z), which is equally oriented and connected with the coordinate system of the cylinder. The distance from the cavity center to the upper boundary of the layer is $y = h$, and the distance to the lower boundary is $y = \tilde{h}$. One needs to find a solution to the Lamé equation. Displacements are set on the layer boundaries $\vec{U}(x,z)|_{y=h} = \vec{f}_h^{\,0}(x,z)$, $\vec{U}(x,z)|_{y=-\tilde{h}} = \vec{f}_{\tilde{h}}^{\,0}(x,z)$, $\vec{U}(\phi,z)|_{\rho=R} = \vec{f}_p^{\,0}(\phi,z)$, there

$$\begin{aligned}\vec{f}_h^{\,0}(x,z) &= U_x^{(h)}\vec{e}_x + U_y^{(h)}\vec{e}_y + U_z^{(h)}\vec{e}_z, \\ \vec{f}_{\tilde{h}}^{\,0}(x,z) &= U_x^{(\tilde{h})}\vec{e}_x + U_y^{(\tilde{h})}\vec{e}_y + U_z^{(\tilde{h})}\vec{e}_z, \\ \vec{f}_p^{\,0}(\phi,z) &= U_\rho^{(p)}\vec{e}_\rho + U_\phi^{(p)}\vec{e}_\phi + U_z^{(p)}\vec{e}_z\end{aligned} \quad (1)$$

—known functions. Among these functions, $U_x^{(h)}$, $U_y^{(h)}$ (Figure 1), $U_z^{(h)}$—where there are periodic functions along the x axis and rapidly decreasing to zero along the z axis. Other functions are assumed to rapidly decrease to zero at large distances from the origin along the z coordinate for the cylinder and along the x and z coordinates for the layer boundaries.

The basic solutions of the Lamé equation are chosen in the form [15]

$$\begin{aligned}\vec{u}_k^{\pm}(x,y,z;\lambda,\mu) &= N_k^{(d)} e^{i(\lambda z + \mu x) \pm \gamma y}; \\ \vec{R}_{k,m}(\rho,\phi,z;\lambda) &= N_k^{(p)} I_m(\lambda\rho) e^{i(\lambda z + m\phi)}; \\ \vec{S}_{k,m}(\rho,\phi,z;\lambda) &= N_k^{(p)}\left[(\text{sign}\lambda)^m K_m(|\lambda|\rho) \cdot e^{i(\lambda z + m\phi)}\right]; k = 1, 2, 3;\end{aligned} \quad (2)$$

$$N_1^{(d)} = \frac{1}{\lambda}\nabla; N_2^{(d)} = \frac{4}{\lambda}(\nu-1)\vec{e}_2^{\,(1)} + \frac{1}{\lambda}\nabla(y\cdot); N_3^{(d)} = \frac{i}{\lambda}\text{rot}\left(\vec{e}_3^{\,(1)}\cdot\right); N_1^{(p)} = \frac{1}{\lambda}\nabla;$$

$$N_2^{(p)} = \frac{1}{\lambda}\left[\nabla\left(\rho\frac{\partial}{\partial\rho}\right) + 4(\nu-1)\left(\nabla - \vec{e}_3^{\,(2)}\frac{\partial}{\partial z}\right)\right]; N_3^{(p)} = \frac{i}{\lambda}\text{rot}\left(\vec{e}_3^{\,(2)}\cdot\right); \gamma = \sqrt{\lambda^2+\mu^2};$$

$$-\infty < \lambda, \mu < \infty, \vec{u}_k^{(-)}, \vec{u}_k^{(+)}$$

where ν—Poisson's ratio; $I_m(x)$, $K_m(x)$—modified Bessel functions; $\vec{R}_{k,m}$, $\vec{S}_{k,m}$, $k = 1, 2, 3$—internal and external solutions of the Lamé equation for the cylinder, respectively; $\vec{u}_k^{(-)}$, $\vec{u}_k^{(+)}$—solutions of the Lamé equation for the layer.

Theorems for the addition of basic solutions in different coordinate systems have the form [15]:

–transition from the solutions $\vec{S}_{k,m}$ of the cylindrical coordinate system to the layer solutions $\vec{u}_k^{(-)}$ (at $y > 0$) and $\vec{u}_k^{(+)}$ (at $y < 0$)

$$\begin{aligned}\vec{S}_{k,m}\left(\rho_p,\phi_p,z;\lambda\right) &= \frac{(-i)^m}{2}\int_{-\infty}^{\infty} \omega_{\mp}^m \cdot e^{-i\mu\bar{x}_p \pm \gamma\bar{y}_p} \cdot \vec{u}_k^{(\mp)} \cdot \frac{d\mu}{\gamma}, k = 1,3; \\ \vec{S}_{2,m}\left(\rho_p,\phi_p,z;\lambda\right) &= \frac{(-i)^m}{2}\int_{-\infty}^{\infty} \omega_{\mp}^m \cdot \left(\left(\pm m\cdot\mu - \frac{\lambda^2}{\gamma} \pm \lambda^2 \bar{y}_p\right)\vec{u}_1^{(\mp)} \mp \lambda^2 \vec{u}_2^{(\mp)} \pm \right. \\ &\left. \pm 4\mu(1-\nu)\vec{u}_3^{(\mp)}\right) \cdot \frac{e^{-i\mu\bar{x}_p \pm \gamma\bar{y}_p} d\mu}{\gamma^2},\end{aligned} \quad (3)$$

where $\gamma = \sqrt{\lambda^2+\mu^2}$; $\omega_{\mp}(\lambda,\mu) = \frac{\mu\mp\gamma}{\lambda}$; $m = 0,\pm 1,\pm 2,\ldots$; $\bar{x}_p$ and $\bar{y}_p$—coordinates for the shifted coordinate system (in our case, equal to zero);

–transition from solutions $\vec{u}_k^{(+)}$ and $\vec{u}_k^{(-)}$ of the layer to solutions $\vec{R}_{k,m}$ of the cylinder

$$\vec{u}_k^{(\pm)}(x,y,z) = e^{i\mu\bar{x}_p \pm \gamma\bar{y}_p} \cdot \sum_{m=-\infty}^{\infty} (i\cdot\omega_{\mp})^m \vec{R}_{k,m}, (k = 1,3);$$

$$\vec{u}_2^{(\pm)}(x,y,z) = e^{i\mu\bar{x}_p \pm \gamma\bar{y}_p} \cdot \sum_{m=-\infty}^{\infty} \left[(i \cdot \omega_{\mp})^m \cdot \lambda^{-2} \left(\left(m \cdot \mu + \bar{y}_p \cdot \lambda^2 \right) \cdot \vec{R}_{1,m} \pm \right. \right. \\ \left. \left. \pm \gamma \cdot \vec{R}_{2,m} + 4\mu(1-\nu)\vec{R}_{3,m} \right) \right], \tag{4}$$

where $\vec{R}_{k,m} = \vec{b}_{k,m}(\rho_p, \lambda) \cdot e^{i(m\Phi_p + \lambda z)}$; $\vec{b}_{1,n}(\rho, \lambda) = \vec{e}_\rho \cdot I'_n(\lambda\rho) + i \cdot I_n(\lambda\rho) \cdot \left(\vec{e}_\phi \frac{n}{\lambda\rho} + \vec{e}_z \right)$;
$\vec{b}_{2,n}(\rho,\lambda) = \vec{e}_\rho \cdot [(4\nu - 3) \cdot I'_n(\lambda\rho) + \lambda\rho I''_n(\lambda\rho)] + \vec{e}_\phi i \cdot m \left(I'_n(\lambda\rho) + \frac{4(\sigma-1)}{\lambda\rho} I_n(\lambda\rho) \right) + \vec{e}_z i\lambda\rho I'_n(\lambda\rho)$;
$\vec{b}_{3,n}(\rho,\lambda) = -\left[\vec{e}_\rho \cdot I_n(\lambda\rho) \frac{n}{\lambda\rho} + \vec{e}_\phi \cdot i \cdot I'_n(\lambda\rho) \right]$; $\vec{e}_\rho, \vec{e}_\phi, \vec{e}_z$—orths of the cylindrical coordinate system.

3. Problem Solving and Stress State Research Results

3.1. Creation and Solving a System of Equations

To take into account the specified periodic displacements, an auxiliary problem was solved for a layer without a cavity in the form

$$\vec{U}_0 = \sum_{k=1}^{3} \int_{-\infty}^{\infty} \sum_{n=-\infty}^{\infty} \left(H_{k,n}^{(0)}(\lambda) \cdot \vec{u}_k^{(+)}(x,y,z;\lambda,\mu_n) + \widetilde{H}_{k,n}^{(0)}(\lambda) \cdot \vec{u}_k^{(-)}(x,y,z;\lambda,\mu_n) \right) d\lambda, \tag{5}$$

where $\vec{u}_k^{(+)}(x,y,z;\lambda,\mu)$ and $\vec{u}_k^{(-)}(x,y,z;\lambda,\mu)$—the basic solutions, which are given by formulas (2), and where the unknown functions $H_{k,n}^{(0)}(\lambda)$, $\widetilde{H}_{k,n}^{(0)}(\lambda)$ must be found from the boundary conditions, which are represented by periodic functions.

To take into account the boundary conditions at the lower boundary of the layer, Equation (5), at $y = -\tilde{h}$, is set to zero. At the upper boundary of the layer, the vector (5), at $y = h$, is equal to the given $\vec{f}_h^{\,0}(x,z)$, represented by the integral and the Fourier series

$$\vec{f}_h^{\,0}(x,z) = \int_{-\infty}^{\infty} \sum_{n=-\infty}^{\infty} \vec{c}_h(\lambda,n) e^{i(\mu_n x + \lambda z)} d\lambda$$

where

$$\vec{c}_h(\lambda,n) = \frac{1}{4\pi \cdot \ell} \int_{-\ell}^{\ell} dx \int_{-\infty}^{\infty} \vec{f}_h^{\,0}(x,z) \cdot e^{-i(\mu_n x + \lambda z)} dz; \tag{6}$$

2ℓ—function period; $\mu_n = n\pi/\ell$.
After equating the vector coefficients at $e^{i(\mu_n x + \lambda z)}$, one obtains

$$\sum_{s=1}^{3} H_{k,n}^{(0)}(\lambda) \vec{d}_s^{\,+}(h;\lambda,\mu_n) + \widetilde{H}_{k,n}^{(0)}(\lambda) \vec{d}_s^{\,-}(h;\lambda,\mu_n) = \vec{c}_h(\lambda,n) \\ \sum_{s=1}^{3} H_{k,n}^{(0)}(\lambda) \vec{d}_s^{\,+}(-\tilde{h};\lambda,\mu_n) + \widetilde{H}_{k,n}^{(0)}(\lambda) \vec{d}_s^{\,-}(-\tilde{h};\lambda,\mu_n) = 0. \tag{7}$$

Equation (7) is projected on the coordinate axis (equalized projections with base vectors $\vec{e}_x, \vec{e}_y, \vec{e}_z$) and expressed as $H_{k,n}^{(0)}(\lambda)$ and $\widetilde{H}_{k,n}^{(0)}(\lambda)$

$$H_{j,n}^{(0)}(\lambda) = \sum_{k=1}^{3} \frac{A_{k,j}}{D} \left(\vec{c}_h(\lambda,n) \right) \cdot \vec{e}_k ; \quad \widetilde{H}_{j,n}^{(0)}(\lambda) = \sum_{k=1}^{3} \frac{A_{k,j+3}}{D} \left(\vec{c}_h(\lambda,n) \right) \cdot \vec{e}_k,$$

where $\vec{e}_k$, $k = 1, 2, 3$—basis vectors of the Cartesian coordinate system; $j = 1, 2, 3$; $A_{1..6,1..6}$—algebraic complement of the system of equations; and D—the determinant of the system of equations.

After determining the unknowns $H_{k,n}^{(0)}(\lambda)$ and $\widetilde{H}_{k,n}^{(0)}(\lambda)$ by using transition formulas (4), expression (7) is rewritten in the cylindrical coordinate system through basic solutions $\vec{R}_{k,m}$ and found displacements at the place where the surface of the cavity is geometrically located. When releasing from the integral over λ and $e^{i(\lambda z + m\phi)}$, one obtains

$$\vec{h}_m^{(0)}(\lambda) = \sum_{n=-\infty}^{\infty} \left[\sum_{s=1}^{3} \left(r_{n,j}(R; m, \lambda) \cdot f_{k,n}^m(\lambda, \mu_n) \cdot \sum_{k=1}^{3} \frac{A_{k,s}}{D} \left(\vec{c}_h(\lambda, n) \right) \right) + \sum_{s=1}^{3} \left(r_{n,j}(R; m, \lambda) \cdot \widetilde{f}_{k,n}^m(\lambda, \mu_n) \cdot \sum_{k=1}^{3} \frac{A_{k,s+3}}{D} \left(\vec{c}_h(\lambda, n) \right) \right) \right], \quad (8)$$

where

$$r_{1,1}(\rho; m, \lambda) = \frac{I_{m+1}(\lambda\rho)}{2} + \frac{I_{m-1}(\lambda\rho)}{2}; \quad r_{1,2}(\rho; m, \lambda) = \frac{i \cdot m \cdot I_m(\lambda\rho)}{\lambda\rho};$$
$$r_{1,3}(\rho; m, \lambda) = i \cdot I_m(\lambda\rho);$$
$$r_{2,1}(\rho; m, \lambda) = \frac{\lambda\rho(I_{m+2}(\lambda\rho) + 2I_m(\lambda\rho) + I_{m-2}(\lambda\rho))}{4} + \left(2\nu - \frac{3}{2}\right) \cdot (I_{m+1}(\lambda\rho) + I_{m-1}(\lambda\rho));$$
$$r_{2,2}(\rho; m, \lambda) = im \cdot \left(\frac{(4\nu - 4)I_m(\lambda\rho)}{\lambda\rho} + \frac{I_{m+1}(\lambda\rho)}{2} + \frac{I_{m-1}(\lambda\rho)}{2} \right); \quad r_{3,1}(\rho; m, \lambda) = -\frac{m \cdot I_m(\lambda\rho)}{\lambda\rho};$$
$$r_{3,2}(\rho; m, \lambda) = -\frac{i}{2} \cdot (I_{m+1}(\lambda\rho) + I_{m-1}(\lambda\rho)), \quad r_{3,3}(\rho; m, \lambda) = 0;$$

$$f_{k,n}^m(\lambda, \mu) = (i \cdot \omega_-(\lambda, \mu))^m \cdot \begin{pmatrix} 1 & 0 & 0 \\ \frac{m\mu}{\lambda^2} & \frac{\gamma}{\lambda^2} & \frac{4\mu(1-\nu)}{\lambda^2} \\ 0 & 0 & 1 \end{pmatrix};$$

$$\widetilde{f}_{k,n}^m(\lambda, \mu) = (i \cdot \omega_+(\lambda, \mu))^m \cdot \begin{pmatrix} 1 & 0 & 0 \\ \frac{m\mu}{\lambda^2} & -\frac{\gamma}{\lambda^2} & \frac{4\mu(1-\nu)}{\lambda^2} \\ 0 & 0 & 1 \end{pmatrix};$$

$\vec{c}_h(\lambda, n)$—presented in formula (6).

After finding the imprint of the periodic function at the location of the cavity (8), the main problem of the generalized Fourier method is solved, the solution of which is given in the form

$$\vec{U} = \vec{U}_0 + \vec{U}_1,$$

where $\vec{U}_0$—presented in formula (5);

$$\vec{U}_1 = \sum_{k=1}^{3} \int_{-\infty}^{\infty} \sum_{m=-\infty}^{\infty} B_{k,m}(\lambda) \cdot \vec{S}_{k,m}(\rho, \phi, z; \lambda) d\lambda + \sum_{k=1}^{3} \int_{-\infty}^{\infty} \int_{-\infty}^{\infty} \left(H_k(\lambda, \mu) \cdot \vec{u}_k^{(+)}(x, y, z; \lambda, \mu) + \widetilde{H}_k(\lambda, \mu) \cdot \vec{u}_k^{(-)}(x, y, z; \lambda, \mu) \right) d\mu d\lambda; \quad (9)$$

$\vec{S}_{k,m}(\rho, \phi, z; \lambda)$, $\vec{u}_k^{(+)}(x, y, z; \lambda, \mu)$ and $\vec{u}_k^{(-)}(x, y, z; \lambda, \mu)$—the basic solutions given by formula (2), and the unknown functions $H_k(\lambda, \mu)$, $\widetilde{H}_k(\lambda, \mu)$ and $B_{k,m}(\lambda)$ must be determined from the boundary conditions (1), taking into account the additional function (8) with the opposite sign on the cavity surface.

Finding the unknowns $H_k(\lambda, \mu)$, $\widetilde{H}_k(\lambda, \mu)$ and $B_{k,m}(\lambda)$ was carried out as in [23].

That is, in order to fulfill the boundary conditions on the lower boundary of the layer $y = -\widetilde{h}$, the vectors $\vec{S}_{k,m}$ in (9) are rewritten in the Cartesian coordinate system using the transition formulas (3) through the basic solutions $\vec{u}_k^{(+)}$. The resulting vectors are equated, at y = $-\tilde{h}$, given by $\vec{f}_{\tilde{h}}^{0}(x,z)$, represented by the double Fourier integral

$$\vec{f}_{\tilde{h}}^{0}(x,z) = \int\limits_{-\infty}^{\infty}\int\limits_{-\infty}^{\infty} \vec{c}_{\tilde{h}}(\lambda,\mu) e^{i(\lambda z + \mu x)} d\lambda d\mu.$$

At the upper boundary of the layer, the boundary conditions are taken into account in Equation (5), so Equation (9) equals zero at y = h. At the same time, the vectors $\vec{S}_{k,m}$ in (9) are rewritten in the Cartesian coordinate system through the basic solutions $\vec{u}_k^{(-)}$ using the transition Formulas (3).

The functions $H_k(\lambda,\mu)$ and $\tilde{H}_k(\lambda,\mu)$ expressed through $B_{k,m}(\lambda)$ are found from these equations.

To fulfill the boundary conditions on the cylinder ρ = R, the right part of (9) is rewritten in the cylindrical coordinate system through the base solutions $\vec{R}_{k,m}$, $\vec{S}_{k,m}$ using the transition formulas (4). The resulting vector is equated to $\vec{h}_m(\lambda) = \vec{h}_m^{(1)}(\lambda) - \vec{h}_m^{(0)}(\lambda)$, where $\vec{h}_m^{(1)}(\lambda)$ is the given function $\vec{f}_p^{0}(\phi,z)$ (1), represented by an integral and Fourier series; $\vec{h}_m^{(0)}(\lambda)$ is the density of the integral image (8).

The previously found functions $H_k(\lambda,\mu)$ and $\tilde{H}_k(\lambda,\mu)$ through $B_{k,m}(\lambda)$ are excluded from the resulting system of equations. As a result, a set of three systems of linear algebraic equations was obtained to determine the unknown $B_{k,m}(\lambda)$.

The functions $B_{k,m}(\lambda)$, found from the infinite system of equations, are substituted into the expressions for $H_k(\lambda,\mu)$ and $\tilde{H}_k(\lambda,\mu)$. This determined all the unknown problems.

3.2. Numerical Analysis of the Stress State of the Layer

The layer properties (fine-grained concrete with quartz aggregate) were taken: modulus of elasticity E = 3.25·10^4 MPa and Poisson's ratio ν = 0.16 [29]. The radius of the cylindrical cavity is R = 100 mm. The upper and lower boundaries of the layer (Figure 1) are located relative to the center of the cavity at a distance of h = $\tilde{h}$ = 200 mm.

At the upper boundary of the layer along the x-axis, a periodic function of displacements is set in the form

$$U_y^{(h)}(x,z) = \begin{cases} -\left(1-\frac{x}{2}\right)\cdot\left(10^4 \cdot (z^2 + 10^2)^{-2}\right), & 0 \le |x| \le 2 \\ 0, & 2 < |x| \le \pi \end{cases},$$

$U_x^{(h)} = U_z^{(h)} = 0$, on the lower boundary, $U_x^{(\tilde{h})} = U_y^{(\tilde{h})} = U_z^{(\tilde{h})} = 0$, and on the surface of the cavity displacement, $U_\rho^{(p)} = U_\phi^{(p)} = U_z^{(p)} = 0$. The representation of the function $U_y^{(h)}(x,z)$ through the Fourier series along the x-axis and the Fourier integral along the z-axis has the form

$$f(x;\lambda,n) = -\left[\frac{2}{\pi}\left(\frac{1}{2} + \sum_{n=1}^{\infty}\left(\frac{\sin n}{n}\right)^2 \cos nx\right)\right] \cdot \left[2,5 e^{-|\lambda|10}(|\lambda|10 + 1)\right].$$

The infinite system of equations was reduced to a finite one by the parameters m = 8 and n = 35. The accuracy of meeting the boundary conditions for the specified values of the geometric parameters is 10^{-3} for values equal to zero.

Stresses were determined along the z axis (Figure 2, line a), on the surface of the cylindrical cavity (Figure 2, line b), on the isthmuses (Figure 2, lines c and d), as well as on the upper boundary of the layer along the x axis (Figure 2, line k).

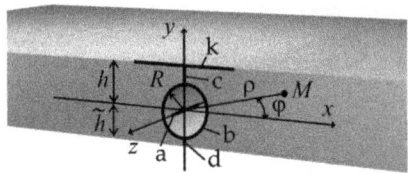

Figure 2. The areas where a stresses was determined.

Figure 3a represents the stress state at the upper boundary of the layer along the z axis, at $x = 0$ (Figure 2, line a). Figure 3b represents the stress state on the surface of the cavity (Figure 2, line b)

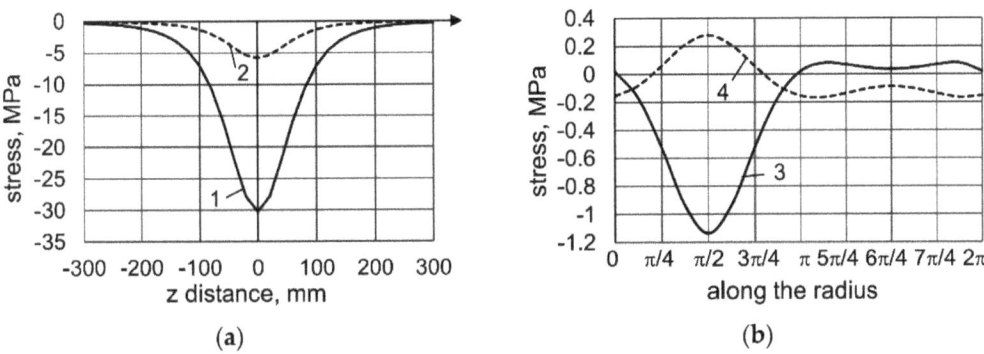

Figure 3. Stressed state: (**a**) at the upper boundary of the layer along the z axis at $x = 0$; (**b**) on the surface of a cylindrical cavity at $z = 0$; $1-\sigma_y$; $2-\sigma_x$; $3-\sigma_\rho$; $4-\sigma_\phi$.

The stresses σ_x (Figure 3a, line 2) and σ_z at the upper boundary of the layer along the z axis coincide. The highest stress values are realized in the section $z = 0$ and are equal to $\sigma_y = -30.2475$ MPa, $\sigma_x = -5.7514$ MPa.

On the surface of the cavity (Figure 2, line b), the stresses decrease significantly (Figure 3b). The highest stress values σ_ρ, the maximum stress values $\sigma_\rho = -1.1359$ MPa and $\sigma_\phi = 0.2793$ MPa are observed in the upper part of the cavity ($\pi/2$).

The stresses at the isthmuses between the cavity and the layer boundaries at $z = 0$ (Figure 2, lines c and d) are represented in Figure 4a,b.

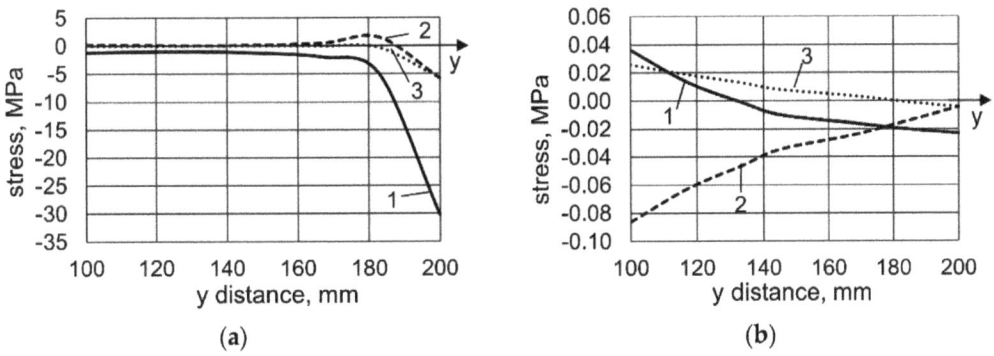

Figure 4. The stress state on the isthmus between the cavity and: (**a**) the upper boundary of the layer; (**b**) the lower boundary of the layer; $1-\sigma_y$; $2-\sigma_x$; $3-\sigma_z$.

The stresses along the isthmus from the upper boundary of the layer to the cavity (Figure 4a) decrease quite quickly, so the influence of the cavity on the stresses is not strongly observed.

The stresses along the isthmus from the cavity to the lower boundary of the layer (Figure 4b) also decrease. In addition, the stresses σ_y and σ_z change signs from "extensive" to "compressive".

Figure 5 represents the stress distribution on the upper boundary of the layer along the x axis at $z = 0$ (Figure 2, line k). The σ_x and σ_z stresses are similar, so the σ_z stresses are not represented.

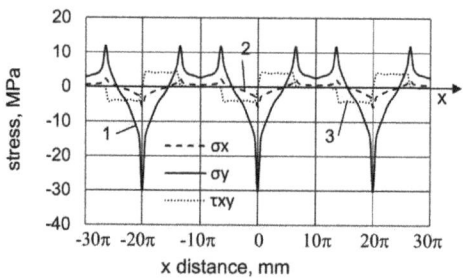

Figure 5. Stress at the upper boundary of the layer along the x axis; 1—σ_y; 2—σ_x; 3—τ_{xy}.

In the locations of the peak "load" of periodic displacement (0, $\pm 10\pi$, $\pm 20\pi$...), maximum negative values of stresses σ_y arise (Figure 5, line 1), which also have a periodic character. Between the "loads", the stresses σ_y become positive. The stresses σ_x, σ_z and τ_{xy} have small values (compared to σ_y), whose maximum values also occur at the peak "load" locations.

4. Discussion

To take into account the periodic load, the solution of the additional problem (5) is applied for a layer without a cylindrical cavity, with the distribution of the load in a Fourier series (6) and the definition of displacements in place of the cavity impression (8). Next, the main problem is solved, where these displacements with the opposite sign are taken into account. The total result is the sum of two problems to be solved.

To obtain the stress state, the stress operator is applied for solutions (5) and (9). Using transition Formulas (3) and (4), it became possible to write solutions in one coordinate system and obtain a numerical result.

Unlike the existing papers [17–28], which also use the generalized Fourier method, an additional problem was applied in the presented work. This allows one to solve problems with boundary conditions in the form of functions directed to infinity.

The proposed high-precision solution method makes it possible to calculate the strength of the structure and its details, whose calculation scheme is a layer with a cylindrical cavity and boundary conditions in the form of periodic functions.

When applying the additional problem, one should take into account the limitation: such a technique is impossible in the direction of the z axis, where there is no expansion in a row along the radius of the cavity. However, taking into account the unchanged geometry of the body along this axis, the periodic load along the z axis can be taken into account without applying the proposed additional problem. In this case, the problem can be solved by the method of separation of variables in combination with the generalized Fourier method.

Further development of this direction is necessary for other types of boundary conditions (in stresses and mixed types).

5. Conclusions

1. A new problem of the theory of elasticity for a layer with a longitudinal circular cylindrical cavity with periodic displacements specified at the upper boundary of the layer is solved. To solve the problem, the analytical-numerical generalized Fourier method is applied for a layer with a cylindrical cavity and an additional problem with a Fourier series expansion is applied for a layer without a cylindrical cavity. This application of the methods made it possible to obtain the value of the stress-strain state with a predetermined accuracy for the indicated problem.

2. In the numerical research process, the maximum stresses of the body of the layer were determined. Thus, the highest stress values take place at the upper boundary of the layer in the section $z = 0$ and are equal to $\sigma_y = -30.2475$ MPa and $\sigma_x = -5.7514$ MPa. The maximum stress values on the surface of the cavity are observed in the upper part, whose values reach $\sigma_\rho = -1.1359$ MPa and $\sigma_\phi = 0.2793$ MPa.

Author Contributions: Conceptualization, V.M.; methodology, V.M.; software, V.S.; validation, B.Y., O.S. and V.S.; formal analysis, B.Y.; investigation, V.M.; resources, O.S.; data curation, V.M.; writing—original draft preparation, V.M.; writing—review and editing, V.M. and B.Y.; visualization, O.S.; supervision, V.M.; project administration, V.M. All authors have read and agreed to the published version of the manuscript.

Funding: This research received no external funding.

Data Availability Statement: Not applicable.

Conflicts of Interest: The authors declare no conflict of interest.

References

1. Aitharaju, V.; Aashat, S.; Kia, H.; Satyanarayana, A.; Bogert, P. Progressive Damage Modeling of Notched Composites. NASA Technical Reports Server 2016. Available online: https://ntrs.nasa.gov/archive/nasa/casi.ntrs.nasa.gov/20160012242.pdf (accessed on 4 August 2022).
2. Tekkaya, A.E.; Soyarslan, C. Finite Element Method. In *CIRP Encyclopedia of Production Engineering*; Laperrière, L., Reinhart, G., Eds.; Springer: Berlin/Heidelberg, Germany, 2014; pp. 508–514. [CrossRef]
3. Ghasemi, A.R.; Razavian, I. Measurement of Variation in Fracture Strength and Calculation of Stress Concentration Factor in Composite Laminates with Circular Hole. *J. Solid Mech.* **2012**, *4*, 226–236.
4. Jafari, M.; Bayati Chaleshtari, M.H.; Ardalani, E. Determination of Optimal Parameters for Finite Plates with a Quasi-Square Hole. *J. Solid Mech.* **2018**, *10*, 300–314.
5. Dastjerdi, S.; Yazdanparast, L. New Method for Large Deflection Analysis of an Elliptic Plate Weakened by an Eccentric Circular Hole. *J. Solid Mech.* **2018**, *10*, 561–570.
6. Abolghasemi, S.; Eipakchi, H.R.; Shariati, M. Investigation of Pre-buckling Stress Effect on Buckling Load Determination of Finite Rectangular Plates with Circular Cutout. *J. Solid Mech.* **2018**, *10*, 816–830.
7. Ugrimov, S.; Smetankina, N.; Kravchenko, O.; Yareshchenko, V. Analysis of laminated composites subjected to impact. In Proceedings of the ICTM 2020: Integrated Computer Technologies in Mechanical Engineering-2020, Kharkiv, Ukraine, 19 January 2021. [CrossRef]
8. Rodichev, Y.M.; Smetankina, N.V.; Shupikov, O.M.; Ugrimov, S.V. Stress-strain assessment for laminated aircraft cockpit windows at static and dynamic load. *Strength Mater.* **2018**, *50*, 868–873. [CrossRef]
9. Smetankina, N.; Ugrimov, S.; Kravchenko, I.; Ivchenko, D. Simulating the process of a bird striking a rigid target. In Proceedings of the 2nd International Conference on Design, Simulation, Manufacturing: The Innovation Exchange, DSMIE-2019, Lutsk, Ukraine, 11–14 June 2019. [CrossRef]
10. Smetankina, N.; Kravchenko, I.; Merculov, V.; Ivchenko, D.; Malykhina, A. Modelling of Bird Strike on an Aircraft Glazing. In Proceedings of the Conference "Integrated Computer Technologies in Mechanical Engineering 2020 – Synergetic Engineering", Kharkiv, Ukraine, 28–30 October 2020; Volume 1113, pp. 289–297. [CrossRef]
11. Guz', A.N.; Kubenko, V.D.; Cherevko, M.A. *Difraktsiya Uprugikh voln*; Naukova dumka: Kyiv, Ukraine, 1978; pp. 18–308.
12. Grinchenko, V.T.; Meleshko, V.V. *Garmonicheskiye Kolebaniya i Volny v Uprugikh Telakh*; Naukova Dumka: Kyiv, Ukraine, 1981; pp. 15–284.
13. Grinchenko, V.T.; Ulitko, A.F. An exact solution of the problem of stress distribution close to a circular hole in an elastic layer. *Sov. Appl. Mech.* **1968**, *4*, 31–37. [CrossRef]

14. Volchkov, V.V.; Vukolov, D.S.; Storozhev, V.I. Difraktsiya voln sdviga na vnutrennikh tunnel'nykh tsilindricheskikh neodnorodnostyakh v vide polosti i vklyucheniya v uprugom sloye so svobodnymi granyami. *Mekhanika Tverdogo Tela: Mezhvedomstvenny Sbornik Nauchnyh Trudov* **2016**, *46*, 119–133.
15. Nikolayev, A.G.; Protsenko, V.S. *Obobshchennyy metod Fur'ye v Prostranstvennykh Zadachakh Teorii Uprugosti*; Nats. Aerokosm. Universitet im. N.Ye. Zhukovskogo «KHAI»: Khar'kov, Ukraine, 2011; pp. 10–344.
16. Ukrayinets, N.; Murahovska, O.; Prokhorova, O. Solving a one mixed problem in elasticity theory for half-space with a cylindrical cavity by the generalized fourier method. *East.-Eur. J. Enterp. Technol.* **2021**, *2*, 48–57. [CrossRef]
17. Nikolaev, A.G.; Tanchik, E.A. Stresses in an Infinite Circular Cylinder with Four Cylindrical Cavities. *J. Math. Sci.* **2016**, *217*, 299–311. [CrossRef]
18. Nikolaev, A.G.; Tanchik, E.A. The first boundary-value problem of the elasticity theory for a cylinder with N cylindrical cavities. *Numer. Anal. Appl.* **2015**, *8*, 148–158. [CrossRef]
19. Nikolaev, A.G.; Tanchik, E.A. Stresses in an elastic cylinder with cylindrical cavities forming a hexagonal structure. *J. Appl. Mech. Tech. Phys.* **2016**, *57*, 1141–1149. [CrossRef]
20. Nikolaev, A.G.; Tanchik, E.A. Model of the Stress State of a Unidirectional Composite with Cylindrical Fibers Forming a Tetragonal Structure. *Mech. Compos. Mater.* **2016**, *52*, 177–188. [CrossRef]
21. Protsenko, V.S.; Popova, N.A. Vtoraya osnovnaya krayevaya zadacha teorii uprugosti dlya poluprostranstva s krugovoy tsilindricheskoy polost'yu. *Dopovidi NAN Ukr.* **2004**, *12*, 52–58.
22. Protsenko, V.; Miroshnikov, V. Investigating a problem from the theory of elasticity for a half-space with cylindrical cavities for which boundary conditions of contact type are assigned. *East.-Eur. J. Enterp. Technol.* **2018**, *4*, 43–50. [CrossRef]
23. Miroshnikov, V.Y. The study of the second main problem of the theory of elasticity for a layer with a cylindrical cavity. *Strength Mater. Theory Struct.* **2019**, *102*, 77–90. [CrossRef]
24. Miroshnikov, V.; Denysova, T.; Protsenko, V. The study of the first main problem of the theory of elasticity for a layer with a cylindrical cavity. *Strength Mater. Theory Struct.* **2019**, *103*, 208–218. [CrossRef]
25. Miroshnikov, V.Y. Stress State of an Elastic Layer with a Cylindrical Cavity on a Rigid Foundation. *Int. Appl. Mech.* **2020**, *56*, 372–381. [CrossRef]
26. Miroshnikov, V. Investigation of the stress state of a composite in the form of a layer and a half space with a longitudinal cylindrical cavity at stresses given on boundary surfaces. *J. Mech. Eng.* **2019**, *22*, 24–31. [CrossRef]
27. Miroshnikov, V.Y.; Medvedeva, A.V.; Oleshkevich, S.V. Determination of the Stress State of the Layer with a Cylindrical Elastic Inclusion. *Mater. Sci. Forum* **2019**, *968*, 413–420. [CrossRef]
28. Miroshnikov, V.; Savin, O.; Younis, B.; Nikichanov, V. Solution of the problem of the theory of elasticity and analysis of the stress state of a fibrous composite layer under the action of transverse compressive forces. *East.-Eur. J. Enterp. Technol.* **2022**, *4*, 23–30. [CrossRef]
29. Bambura, A.; Barashikov, A.; Golyshev, O.; Krivosheev, P. *Derzhavni Budivel'ni Normy Ukrayiny. Betonni ta Zalizobetonni Konstruktsiyi Osnovni Polozhennya (DBN V.2.6-98:2008)*; Minrehionbud Ukrayiny: Kyiv, Ukraine, 2011; p. 71. Available online: https://dbn.co.ua/load/normativy/dbn/1-1-0-792 (accessed on 4 August 2022).

Article

Signal Processing Algorithm for Monopulse Noise Noncoherent Wideband Helicopter Altitude Radar

Valeriy Volosyuk [1], Volodymyr Pavlikov [1,*], Simeon Zhyla [1], Eduard Tserne [1,*], Oleksii Odokiienko [1], Andrii Humennyi [2], Anatoliy Popov [1] and Oleh Uruskiy [3]

[1] Aerospace Radio-Electronic Systems Department, National Aerospace University "Kharkiv Aviation Institute", 61070 Kharkiv, Ukraine
[2] National Aerospace University "Kharkiv Aviation Institute", 61070 Kharkiv, Ukraine
[3] Central Scientific Research Institute of Armament and Military Equipment of the Armed Forces of Ukraine, 03049 Kyiv, Ukraine
* Correspondence: v.pavlikov@khai.edu (V.P.); e.tserne@khai.edu (E.T.)

Abstract: Radio altimeters are an important component of modern helicopter on-board systems. These devices currently involve the use of narrowband deterministic signals, that limits their potential technical characteristics. Given the significant breakthrough in the development of wideband and ultra-wideband radio electronics, it is promising to create on-board radio complexes capable of obtaining the necessary information using wideband stochastic signals. At the same time, when developing such complexes, it is important to use optimal synthesis methods for radio systems, which will allow optimal signal processing algorithms and potential accuracy parameters to be obtained. In this work, the algorithm to measure flight altitude for a helicopter or an unmanned aerial vehicle based on the processing of wideband and ultra-wideband pulsed stochastic signals is synthesized for the first time by the maximum-likelihood method. When formulating the problem, the mathematical model of the signal and observation is specified, and their statistical characteristics are investigated. The peculiarity of the synthesis task is the use of a noise pulse transmitter, which implements the function of an underlying surface illuminator, as well as considering the signal structure destruction during its radiation, propagation, and reflection. This signal shape destruction makes it impossible to synthesize a radar with internally coherent processing when working on one receiving antenna. In accordance with the synthesized algorithm, a simulation model of a pulsed radar with a stochastic probing signal has been developed and the results of its modeling are presented.

Keywords: broadband stochastic signals; radar altimeter; helicopter radar; optimal signal processing algorithm

1. Introduction

Motivation: Modern trends in the development of all types of manned and unmanned aircraft indicate the importance of issues relating to ensuring flight safety and improving autopilot capabilities [1–3]. Solving these issues requires a comprehensive approach and the simultaneous use of a large number of on-board systems for continuous monitoring of both the condition of the aircraft itself (its speed, acceleration, roll, altitude, coordinates, etc.) and the current condition of the environment (wind speed and direction, pressure, etc). The information received from all aircraft sensors is mostly sent to a single on-board complex, processed, and further used by the pilot to make decisions and the automatic control system to generate the necessary control signals depending on the current task and the determined condition of the aircraft.

In terms of the complexity of ensuring flight safety and implementing autopilot systems, helicopters stand out among many other types of aircraft. Therefore, their use involves the possibility of flights in a wide range of speeds, heights, and directions. The information received from the on-board complex is used to make decisions about maneuvering, solving navigation problems, and avoiding dangerous situations. An important element of such a complex is an altimeter, which allows you to measure the current aircraft altitude above the ground surface with high accuracy and with a high degree of reliability.

State of the Art: Today, there is a large number of altimeters, ranging from barometric [4], to gamma ray devices. They can be based on different principles of altitude measurement, and even similar systems may involve the use of different signal types with different frequencies and work algorithms. For example, laser radiation is used and processed in lidar systems [5–7], there are also systems that use telecommunication signals [8,9], nonlinear algorithms for a movement trajectory processing [10–12], etc. Traditionally, the most reliable are radio altimeters [13–15], which work in the radio range of waves, which allows one to receive the necessary information regardless of the current time of day, weather conditions, and are completely autonomous. Such systems can be implemented according to different schemes, but they all have some common features, such as the use of pulsed or continuous deterministic narrowband probing signals, which limits their potential technical characteristics. At the same time, wideband and ultra-wideband measuring systems with noise signals [16–18], are currently of considerable interest, because in recent years the radio element base has reached a level sufficient for their implementation. Thus, advanced developers in the radio electronics field currently offer a wide selection of high-speed analog-to-digital converters [19,20], wideband amplifiers [21], antennas [22], etc.

The use of stochastic wideband signals opens up new possibilities in the aerospace-based radio vision systems design. Such systems can provide much better measurement accuracy along with a high degree of protection against extraneous radiation. The main disadvantage of such radars is the significant complexity of implementing a coherent mode of signal processing, which can be achieved only at small distances of several tens of meters [23], even with the use of correction models [24], for the atmosphere influence on signal propagation.

Therefore, the creation of on-board radio complexes capable of obtaining the necessary information using stochastic signals is an urgent task today [25–27]. Radio complexes that will be placed on board the helicopter are no exception. It should be noted that it is advisable to search for the algorithms of such systems using the methods of optimal synthesis [28], of radio systems, which allow to obtain both the optimal algorithm for estimating the desired parameter and the potential accuracy of this estimate.

Objectives: This work is aimed at solving the problem of statistical synthesis of the stochastic radio-signal processing algorithm for measuring the flight height of a helicopter or an unmanned aerial vehicle.

2. Materials and Methods

In Figure 1 a helicopter is shown, that moves at a speed $\vec{V}'$ and is at a height h above the ground surface at a moment in time t. The antenna of the radio altimeter is marked A, and the area of radiation is marked by D'. The area of the underlying surface, which is irradiated by the radiation pattern, is marked by D. Radius vectors $\vec{r}'$ and $\vec{r}$ denote the positions of elementary areas within the radiation area D' and the radiated area D. At the same time, the projections of these radius vectors beginning on the underlying surface coincide. The projection of the velocity $\vec{V}'$ onto the underlying surface is denoted by the vector $\vec{V}$. The project line of the path to the underlying surface is marked as S.

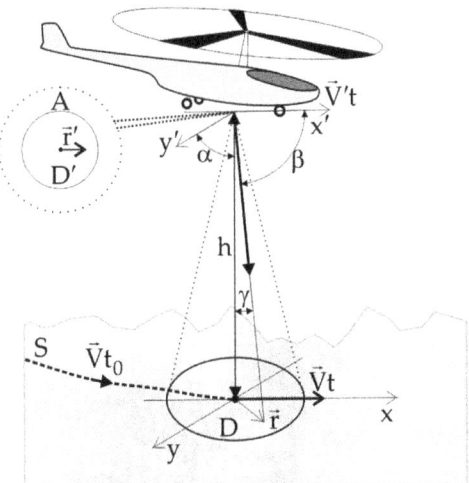

Figure 1. Physical parameters and geometric relationships used in the synthesis of the helicopter altitude measurement algorithm.

Helicopter altimeters often use two antennas [29,30], one for signal radiation and one for reception. The use of one antenna in the modes of operation for radiation and reception of signals is due to the fact that the intended radar will work with pulsed stochastic signals.

The purpose of the work is to synthesize an algorithm for processing wideband stochastic signals for measuring the true altitude of a helicopter.

For the signal processing algorithm synthesis, we will develop a model of a pulsed (for the implementation of a radar that will work with one antenna) stochastic signal and determine its statistical characteristics. Here and further, considering the complexity of mathematical explanations, we will use two approaches to describe the signals at once, temporal and spectral.

The following signal model has been developed:

$$s(t) = \mathcal{F}_f^{-1}\left\{ \Pi(f) \int_{-\infty}^{\infty} \dot{P}(f-f_1)\dot{N}(f_1) df_1 \right\} = \int_{-\infty}^{\infty} \Pi(f) \int_{-\infty}^{\infty} \dot{P}(f-f_1)\dot{N}(f_1) df_1\, e^{j2\pi f t} df = \int_{-\infty}^{\infty} \eta(t-t') P(t') n(t') dt', \quad (1)$$

where $\mathcal{F}_f^{-1}\{\cdot\}$ is operator notation of the inverse Fourier transform with integration over a variable f; $\dot{N}(f) = \mathcal{F}_f^{-1}\{n(t)\}$ is the complex stochastic spectral density of the radiated signal amplitude, ($n(t)$ is a white Gaussian noise with zero mean, which is used as the filling of the radiated radio pulse); $\Pi(f) = \mathcal{F}_f^{-1}\{\eta(t)\}$ is the spectrum of the radio pulse envelope ($P(t)$ is radio pulse envelope); f is frequency; t is time; and τ is pulse duration in time. In the problem being solved, it is assumed that the bandwidth satisfies the condition of wideband or ultra-wideband.

Today, Equation (1) signal type can be generated by the existing radio element base [31,32].

It is assumed that the stochastic spectral density of the useful signal is a Gaussian process with zero mean, that is $\langle \dot{N}(f) \rangle = 0$ or $\langle n(t) \rangle = 0$, in the time domain. Here and further, parentheses $\langle \cdot \rangle$ denote statistical averaging. The correlation function of this process is delta-correlated by frequency $\langle \dot{N}(f_1)\dot{N}^*(f_2) \rangle = (N/2)\,\delta(f_1 - f_2)$ or in the time domain $\langle n(t_1)n(t_2) \rangle = (N/2)\,\delta(t_1 - t_2)$, where $\delta(\cdot)$ is a delta function. Here N is the power spectral density of the probing signal.

Often, only one of these records is used in professional literature. The correctness of both entries can be easily proven by the following calculation:

$$\left\langle \dot{N}(f_1)\dot{N}^*(f_2) \right\rangle = \left\langle \int\limits_{-\infty}^{\infty} n(t)e^{-j2\pi f_1 t}dt \int\limits_{-\infty}^{\infty} n(t) \times e^{j2\pi f_2 t}dt \right\rangle$$

$$= \int\limits_{-\infty}^{\infty}\int\limits_{-\infty}^{\infty} \underbrace{\langle n(t_1)n(t_2)\rangle}_{\frac{N}{2}\delta(t_1-t_2)} e^{-j2\pi(f_1 t_1-f_2 t_2)}dt_1 dt_2 = \frac{N}{2}\int\limits_{-\infty}^{\infty} e^{-j2\pi(f_1-f_2)t}dt = \underbrace{\frac{N}{2}\delta(f_1-f_2)}_{\delta(f_1-f_2)}. \quad (2)$$

With the above limitations, the full information about the signal is contained in its correlation function or power spectral density. Signal correlation function of Equation (1) has the following form:

$$R_s(t_1,t_2) = \langle s(t_1)s(t_2) \rangle = \int\limits_{-\infty}^{\infty}\int\limits_{-\infty}^{\infty} \eta(t_1-t')\eta(t_2-t'')P(t')P(t'')$$

$$\times \underbrace{\langle n(t')n(t'') \rangle}_{\frac{N}{2}\delta(t'-t'')} dt'dt'' = \frac{N}{2}\int\limits_{-\infty}^{\infty} \eta(t_1-t)\eta(t_2-t)P^2(t)dt. \quad (3)$$

The correlation function (3) depends on (t_1, t_2) and not on the difference $t_1 - t_2$, i.e., it describes a non-stationary signal.

The signal power spectral density can be found according to the generalized Wiener–Khinchin theorem. To do this, we perform the Fourier transformation of Equation (3):

$$G_s(t_1,f) = \mathcal{F}_\tau\left\{ R_s(t_1,t_2)|_{t_2=t_1-\tau} \right\} = \frac{N}{2}\mathcal{F}_\tau\left\{ \int\limits_{-\infty}^{\infty} \eta(t_1-t)\eta(t_1-\tau-t)P^2(t)dt \right\} =$$

$$= \frac{N}{2}\dot{\Pi}^*(j2\pi f) \int\limits_{-\infty}^{\infty}\int\limits_{-\infty}^{\infty} \dot{\Pi}(j2\pi[f_2+f_3+f])\,\dot{P}(j2\pi f_2)e^{j2\pi f_2 t_1}\,df_2\dot{P}(j2\pi f_3)\,e^{j2\pi f_3 t_1}\,df_3. \quad (4)$$

For the signal processing algorithm synthesis, it is important to write down the observation equation model. At the same time, various variants of the input path implementation are considered [33,34], which impose restrictions on the observation equation form, as well as the geometry of the problem. It is known from the statistical theory basics [28], that the observation equation model in this case can be described quite accurately by an additive mixture of the useful signal reflected by the underlying surface and noise [35]:

$$u(t) = \int\limits_{D} s_s\left(t,\vec{r}\right)d\vec{r} + n_r(t) + n(t) \quad (5)$$

where $s_s\left(t,\vec{r}\right)$ is a signal reflected by an elementary section of the underlying surface with the center coordinates determined by the end of the radius vector $\vec{r}$; $n_r(t) + n(t)$ is a noise additive taking into account the limitation of the receiver working bandwidth and white noise. We neglect the Doppler frequency in Equation (5), because in the considered geometry of the problem it will be close to zero [36].

Mathematical models of signal and noises can be presented as follows:

$$s_s\left(t,\vec{r}\right) = \int\limits_{-\infty}^{\infty} \dot{\Pi}(f)\dot{F}\left(\vec{r},f\right)\dot{G}\left(\vec{r},f\right) \times \int\limits_{-\infty}^{\infty} \dot{P}(f-f_1)\dot{N}(f_1)\,df_1\,e^{j2\pi f(t-t_d(\vec{r}))}df = \int\limits_{-\infty}^{\infty} n(t')P(t') \int\limits_{-\infty}^{\infty} \dot{\Pi}(f)\dot{F}\left(\vec{r},f\right)\dot{G}\left(\vec{r},f\right) \times$$

$$\times e^{j2\pi f(t-t'-t_d(\vec{r}))}df\,dt' = \int\limits_{-\infty}^{\infty} g\left(\vec{r},t^\sim\right) \int\limits_{-\infty}^{\infty} \eta(t'') \int\limits_{-\infty}^{\infty} n(t')P(t')\,f_\sigma\left(\vec{r},t-t'-t''-t^\sim-t_d\left(\vec{r}\right)\right)dt'\,dt''\,dt^\sim, \quad (6)$$

$$n_r(t) = \int_{-\infty}^{\infty} \eta(t-t_1)n(t_1)\,dt_1 = \int_{-\infty}^{\infty} \dot{\Pi}(f)\dot{N}_n(f)\,e^{j2\pi ft}df,$$

$$n(t) = \int_{-\infty}^{\infty} \dot{N}_0(f)e^{j2\pi ft}df, \tag{7}$$

where $\dot{F}(\vec{r},f)$ is the underlying surface complex reflection coefficient (we consider it a random process [28]), which depends on the frequency [37], when working with wideband signals; $f_\sigma(\cdot) = \mathcal{F}_f^{-1}\{F(\vec{r},f)\}$; $\dot{N}_n(f)$ and $\dot{N}_0(f)$ are spectral densities of the complex amplitude of receiver noise and white noise; $\dot{G}(\vec{r},f)$ is the radiation pattern of the antenna A as a function of frequency, recalculated to the underlying surface elements with coordinates $\vec{r} \in D$ (the beginning of the radius vector $\vec{r}$ is located in the projection of the antenna phase center onto the underlying surface); and $t_d(\vec{r}) = 2c^{-1}\sqrt{h^2 + |\vec{r}|^2}$ is the delay time for the signal to propagate from the phase center of the antenna to the elemental section of the underlying surface with the coordinates determined by the vector $\vec{r}$ end and in the reverse direction.

We ignore the signal attenuation coefficient during propagation in the atmosphere (for the millimeter range it is advisable [38]. to calculate it taking into account the local properties of the atmosphere), considering that it can be included in the complex radiation pattern.

The complex radiation pattern is related to the amplitude-phase distribution of the field in the antenna aperture by the following formula

$$\dot{G}(\vec{\vartheta},f) = fc^{-1}\int_{D'} \dot{I}(f\vec{r}'c^{-1})e^{-j2\pi f\vec{\vartheta}\vec{r}c^{-1}}d\vec{r}' \tag{8}$$

and is recalculated to the underlying surface coordinates, taking into account Figure 1, as follows

$$\dot{G}(\vec{r},f) = \dot{G}\left(\left[h\frac{\cos\alpha}{\cos\gamma}, h\sqrt{tg^2\gamma - \frac{\cos^2\alpha}{\cos^2\gamma}}\right], f\right) \tag{9}$$

where $\vec{\vartheta} = (\vartheta_x, \vartheta_y, \vartheta_y) = (\cos(\alpha), \cos(\beta), \cos(\gamma))$ are the direction cosines of a unit vector direction of which is characterized by the end of the vector $\vec{r}$; $\vartheta_z = \cos(\gamma) = \sqrt{1 - \vartheta_x^2 - \vartheta_y^2}$; $\alpha \in (0,\pi); \beta \in (0,\pi); \gamma \in (0,\frac{\pi}{2})$ is the angle, that is calculated from the axis $0z'$ directed from the phase center of the radio altimeter antenna in the direction of the normal to the underlying surface elements.

Considering that all processes included in Equation (5) are Gaussian with zero mean, we obtain that the mathematical expectation of the observation $\langle u(t)\rangle = 0$. Considering the mutual uncorrelation of the signal and noise, we write the expression for the correlation function as follows:

$$R_u(t_1,t_2) = \langle u(t_1)u(t_2)\rangle = \iint_{D\,D}\langle s_s(t_1,\vec{r}_1)s_s(t_2,\vec{r}_2)\rangle d\vec{r}_1 d\vec{r}_2 + \\ + \langle n_r(t_1)n_r(t_2)\rangle + \langle n(t_1)n(t_2)\rangle = R_s(t_1,t_2) + R_{nr}(t_1-t_2) + R_n(t_1-t_2). \tag{10}$$

In Equation (10), the partial correlation functions of noises have the following forms:

$$R_{nr}(t_1-t_2) = \langle n_r(t_1)n_r(t_2)\rangle = \int_{-\infty}^{\infty}\int_{-\infty}^{\infty}\eta(t_1-t')\eta(t_2-t'')\langle n(t')n(t'')\rangle dt'\,dt'' = \\ = N_r\int_{-\infty}^{\infty}\eta(t_1-t')\eta(t_2-t')\,dt' = N_r R_\eta(t_1-t_2), \tag{11}$$

$$R_\eta(t_1 - t_2) = R_\eta(\tau) = \int_{-\infty}^{\infty} \eta(t_1 - t')\eta(t_2 - t')\,dt' = \int_{-\infty}^{\infty} \eta(t)\eta(t - \tau)\,dt,$$

$$R_n(t_1 - t_2) = \langle n(t_1)n(t_2) \rangle = N\delta(t_1 - t_2) \tag{12}$$

The expression for the correlation function of the signal can be found through the spectral representation of the signal:

$$\begin{aligned}R_s(t_1, t_2) &= \tfrac{1}{2} \int_D \int_D \int_{-\infty}^{\infty} \int_{-\infty}^{\infty} \dot{G}(\vec{r}_1, f_a)\dot{G}^*(\vec{r}_2, f_b)\langle \dot{F}(\vec{r}_1, f_a)\dot{F}^*(\vec{r}_2, f_b)\rangle \dot{\Pi}(f_a)\dot{\Pi}^*(f_b) \int_{-\infty}^{\infty} \int_{-\infty}^{\infty} \dot{P}(f_a - f_1)\dot{P}^*(f_b - f_2) \times\\ &\times \langle \dot{N}(f_1)\dot{N}^*(f_2)\rangle df_1 df_2\, e^{-j2\pi f_a(t_1 - t_d(\vec{r}_1))} e^{j2\pi f_b(t_2 - t_d(\vec{r}_2))} df_a d\vec{r}_1 df_b d\vec{r}_2 = \tfrac{1}{2}N \int_D \int_D \int_{-\infty}^{\infty} \int_{-\infty}^{\infty} \dot{G}(\vec{r}_1, f_a)\dot{G}^*(\vec{r}_2, f_b) \times\\ &\times \sigma^0(\vec{r}_1 - \vec{r}_2, f_a - f_b) \dot{\Pi}(f_a)\dot{\Pi}^*(f_b) \int_{-\infty}^{\infty} \dot{P}(f_a - f)\dot{P}^*(f_b - f)df\, e^{-j2\pi f_a(t_1 - t_d(\vec{r}_1))} e^{j2\pi f_b(t_2 - t_d(\vec{r}_2))} df_a df_b d\vec{r}_1 d\vec{r}_2.\end{aligned} \tag{13}$$

In Equation (13), it is taken into account that the complex radar cross-section correlation function is associated in space and frequency, i.e., it is written in form $\langle \dot{F}(\vec{r}_1, f_a)\dot{F}^*(\vec{r}_2, f_b)\rangle = \sigma^0(\vec{r}_1 - \vec{r}_2, f_a - f_b)$. Parameter $\sigma^0(\vec{r}_1 - \vec{r}_2, f_a - f_b)$ is the radar cross-section of the underlying surface as a function of frequency and space mismatch. The issue of concretizing the analytical expression for $\sigma^0(\vec{r}_1 - \vec{r}_2, f_a - f_b)$ is quite complex and requires the solution of direct problems of radio physics. In many real situations, it is possible to assume that $\sigma^0(\vec{r}_1 - \vec{r}_2, f_a - f_b) = \sigma^0(f_a - f_b, \vec{r}_1)\delta(\vec{r}_1 - \vec{r}_2)$, i.e., the effective scattering surface is uncorrelated in spatial coordinates. This can happen in practice when a real (non-mirror) underlying surface is irradiated with a millimeter wave range. Then the correlation function (13) can be written as follows

$$\begin{aligned}R_s(t_1, t_2)\big|_{\sigma^0(f_a - f_b)\delta(\vec{r}_1 - \vec{r}_2)} &= \tfrac{1}{2}N \int_D \int_{-\infty}^{\infty} \int_{-\infty}^{\infty} \dot{G}(\vec{r}, f_a)\,G^*(\vec{r}, f_a - \Delta f)\,\sigma^0(\Delta f, \vec{r})\,\dot{\Pi}(f_a)\dot{\Pi}^*(f_a - \Delta f) \times\\ &\times \int_{-\infty}^{\infty} \dot{P}(f_a - f)\dot{P}^*(f_a - \Delta f - f)df\, e^{-j2\pi(\Delta f t_1 + (f_a - \Delta f)\tau)} e^{j2\pi \Delta f t_d(\vec{r})} df_a d\Delta f d\vec{r}.\end{aligned}$$

Based on the obtained formulas, let us write down the final form of the observation correlation function:

$$\begin{aligned}R_u(t_1, t_2) &= R_s(t_1, t_2)\big|_{\sigma^0(f_a - f_b)\delta(\vec{r}_1 - \vec{r}_2)} + R_{nr}(t_1 - t_2) + R_n(t_1 - t_2) = \tfrac{1}{2}N \int_D \int_{-\infty}^{\infty} \int_{-\infty}^{\infty} \dot{G}(\vec{r}, f_a)\dot{G}^*(\vec{r}, f_a - \Delta f) \times\\ &\times \sigma^0(\Delta f, \vec{r})\dot{\Pi}(f_a)\dot{\Pi}^*(f_a - \Delta f) \int_{-\infty}^{\infty} \dot{P}(f_a - f)\dot{P}^*(f_a - \Delta f - f)df\, e^{-j2\pi(\Delta f t_1 + (f_a - \Delta f)\tau)} e^{j2\pi \Delta f t_d(\vec{r})} df_a d\Delta f d\vec{r} +\\ &+ N_r \int_{-\infty}^{\infty} \eta(t)\eta(t - \tau)\,dt + N\delta(\tau).\end{aligned} \tag{14}$$

Equation (14) analysis makes it possible to make an important conclusion. The average value of the observation is zero $\langle u(t) \rangle = 0$, and the correlation function (14) contains information about both the signal delay time and the radio pulse envelop, which are converted into the desired parameter—the range.

To solve the problem of the signal processing algorithm synthesizing we use the maximum likelihood method. In this case, there is a possibility of determining the altitude parameter according to one of the two variants of range estimation; differentiation of the likelihood function by the delay time, or by the radio pulse envelope. Within the scope of this work, the search for the algorithm is carried out by differentiation by delay time.

In this work we write the natural logarithm of the likelihood function in the following form:

$$\ln P\left[u(t)\,\big|\,t_d(\vec{r})\right] = \ln k\left(t_d(\vec{r})\right) - \frac{1}{2}\int_0^T \int_0^T u(t_1)\,W_u\left(t_1, t_2, t_d(\vec{r})\right) u(t_2)\,dt_1 dt_2, \tag{15}$$

where $k\left(t_d\left(\vec{r}\right)\right)$ is a coefficient, the derivative of which depends on the information parameter; $W_u\left(t_1, t_2, t_d\left(\vec{r}\right)\right)$ is a function, which is the inverse of the correlation function (14) and is found from the solution of the inversion equation

$$\int_0^T R_u(t_1, t_2) W_u(t_2, t_3)\, dt_2 = \delta(t_1 - t_3) \qquad (16)$$

When solving the problem, it should be noted the difficulties that naturally arise due to the fact that the underlying surface can be significantly uneven and contain significant differences in height within the area irradiated by the radiation pattern.

3. Results
3.1. Delay Time Estimation Algorithm Synthesis

To obtain the signal processing algorithm, we differentiate Equation (15) by the desired parameter and equate the differentiation result to zero. That is, it is necessary to solve the following equation

$$\frac{\delta \ln P\left[u(t) \mid t_d\left(\vec{r}\right)\right]}{\delta t'_d\left(\vec{r}\right)} = 0 \qquad (17)$$

where $\frac{\delta}{\delta t_d\left(\vec{r}\right)}$ is a functional derivative of the delay time, as a function of the underlying surface coordinates.

Considering that it is possible to obtain the solution of the inversion Equation (16) in an explicit form for a rather limited class of correlation functions, we will perform calculations in the frequency or time–frequency domain [28]. To do this, we will rewrite the correlation function in the frequency (frequency–time) domain. This is due to the fact that all the factors under the integrals are symmetrical in terms of the function's frequency. Let us find the Fourier transformation of the correlation function and write the power spectral density as follows:

$$G_R(f, t_1) = \mathcal{F}_\tau\left\{R_u(t_1, \tau)|_{\sigma^0(\Delta f)\delta(\vec{r}_1 - \vec{r}_2)}\right\} = G_s(f, t_1) + G_{nr}(f) + G_n(f) \qquad (18)$$

where

$$G_s(f, t_1) = \mathcal{F}_\tau\left\{R_s(t_1, \tau)|_{\sigma^0(\Delta f)\delta(\vec{r}_1 - \vec{r}_2)}\right\} = \tfrac{1}{2} N \int_D \int_{-\infty}^{\infty} \dot{G}\left(\vec{r}, \Delta f - f\right) \dot{G}^*\left(\vec{r}, -f\right) \sigma^0\left(\Delta f, \vec{r}\right) \dot{\Pi}(\Delta f - f) \dot{\Pi}^*(-f) \times \\ \times \int_{-\infty}^{\infty} \dot{P}(\Delta f - f - f') \dot{P}^*(-f - f') df'\, e^{-j2\pi \Delta f t_1} e^{j2\pi \Delta f t_d(\vec{r})} d\Delta f\, d\vec{r}, \qquad (19)$$

$$G_{nr}(f) = \mathcal{F}_\tau\{R_{nr}(\tau)\} = N_r\left|\dot{\Pi}(f)\right|^2, \qquad (20)$$

$$G_n(f) = \mathcal{F}_\tau\{R_n(\tau)\} = N \int_{-\infty}^{\infty} \delta(\tau) e^{-j2\pi f \tau} d\tau = N. \qquad (21)$$

The inversion Equation (16) for a non-stationary process in the spectral domain is found in the following form:

$$G_W(t_3, -f) = 2\frac{e^{j2\pi f(t_1 - t_3)}}{G_R(t_1, f)} \qquad (22)$$

The inversion Equation (22) in the frequency domain for the statistical characteristics of non-stationary processes is obtained for the first time. Here, the minus sign at the frequency can be omitted since the autocorrelation function of observation is an even function.

The likelihood equation in the time domain can be written as follows:

$$-\frac{1}{2}\int_0^T\int_0^T \frac{\delta R\left(t_1,t_2,t_d\left(\vec{r}'\right)\right)}{\delta t_d\left(\vec{r}\right)} W\left(t_1,t_2,t_d\left(\vec{r}'\right)\right) dt_1 dt_2 - \frac{1}{2}\int_0^T\int_0^T u(t_1) \frac{\delta W\left(t_1,t_2,t_d\left(\vec{r}'\right)\right)}{\delta t_d\left(\vec{r}\right)} u(t_2) dt_1 dt_2 = 0. \quad (23)$$

To solve this equation, it is necessary to determine the expression for function $W\left(t_1,t_2,t_d\left(\vec{r}'\right)\right)$. Let us rewrite likelihood equation in the spectral domain. For this, the following preliminary calculations are performed:

$$\frac{1}{2}\int_0^T\int_0^T \frac{\delta R\left(t_1,t_2,t_d\left(\vec{r}'\right)\right)}{\delta t_d\left(\vec{r}\right)} \int_{-\infty}^{\infty} G_W\left(t_1,f,t_d\left(\vec{r}'\right)\right) e^{j2\pi f t_2} df dt_1 dt_2 = \frac{1}{2}\int_0^T\int_{-\infty}^{\infty} \frac{\delta \hat{G}_R\left(t_1,f,t_d\left(\vec{r}'\right)\right)}{\delta t_d\left(\vec{r}\right)} G_W\left(t_1,f,t_d\left(\vec{r}'\right)\right) df dt_1,$$

$$\frac{1}{2}\int_0^T\int_0^T u(t_1) \frac{\delta W\left(t_1,t_2,t_d\left(\vec{r}'\right)\right)}{\delta t_d\left(\vec{r}\right)} u(t_2) dt_1 dt_2 = \frac{1}{2}\int_0^T u(t_1) \int_{-\infty}^{\infty} \frac{\delta G_W\left(t_1,f,t_d\left(\vec{r}'\right)\right)}{\delta t_d\left(\vec{r}\right)} \hat{U}^*(j2\pi f) df dt_1,$$

where

$$\hat{U}^*(j2\pi f) = \int_0^T u(t_2) e^{j2\pi f t_2} dt_2$$

and the sign «^» does not denote the true value of the spectrum or spectral density, but its an estimate obtained through the Fourier transformation of a limited observation or correlation function realization.

Now let us write the likelihood equation in the time–frequency domain:

$$-\frac{1}{2}\int_0^T\int_{-\infty}^{\infty} \frac{\delta \hat{G}_R\left(t_1,f,t_d\left(\vec{r}'\right)\right)}{\delta t_d\left(\vec{r}\right)} G_W\left(t_1,f,t_d\left(\vec{r}'\right)\right) df dt_1 - \frac{1}{2}\int_0^T u(t_1) \int_{-\infty}^{\infty} \frac{\delta G_W\left(t_1,f,t_d\left(\vec{r}'\right)\right)}{\delta t_d\left(\vec{r}\right)} \hat{U}^*(j2\pi f) df dt_1 = 0. \quad (24)$$

Substitute Equation (22) into the likelihood Equation (24) and take into account that

$$G_W(t_m,-f) = 2\frac{e^{j2\pi f(t_n-t_m)}}{G_R(t_n,f)}\bigg|_{m=n} = \frac{2}{G_R(t_m,f)}$$

We obtained the following calculation result:

$$-\int_0^T\int_{-\infty}^{\infty} \frac{\delta \hat{G}_R\left(t_1,f,t_d\left(\vec{r}'\right)\right)}{\delta t_d\left(\vec{r}\right)} \frac{1}{G_R\left(t_1,f,t_d\left(\vec{r}'\right)\right)} df dt_1 - \int_0^T u(t_1) \int_{-\infty}^{\infty} \left(\frac{\delta}{\delta t_d\left(\vec{r}\right)} \frac{1}{G_R\left(t_1,f,t_d\left(\vec{r}\right)\right)}\right) \hat{U}^*(j2\pi f) df dt_1 = 0 \quad (25)$$

Next, we find the derivatives that are included in Equation (25). To do this, we use spectral densities (18) and (19). The derivative of the power spectral density (18) has the following form:

$$\frac{\delta \hat{G}_R\left(t_1,f,t_d\left(\vec{r}'\right)\right)}{\delta t_d\left(\vec{r}\right)} = \lim_{\alpha \to 0} \frac{d}{d\alpha} \hat{G}_R\left(t_1,f,t_d\left(\vec{r}'\right) + \alpha\delta\left(t_d\left(\vec{r}'\right) - t_d\left(\vec{r}\right)\right)\right) = j\pi N \int_{-\infty}^{\infty} \Delta f \dot{G}\left(\vec{r}, \Delta f - f''\right) \times$$
$$\times \dot{G}^*\left(\vec{r}, -f''\right) \sigma^0\left(\Delta f, \vec{r}\right) \dot{\Pi}\left(\Delta f - f''\right) \dot{\Pi}^*(-f'') \int_{-\infty}^{\infty} \dot{P}(\Delta f - f'' - f) \dot{P}^*(-f''-f) df e^{-j2\pi \Delta f(t_1 - t_d(\vec{r}))} d\Delta f. \quad (26)$$

The derivative of the power spectral density, which is the inverse of the observation signal power spectral density (taking into account Equation (26)), can be written as follows:

$$\frac{\delta}{\delta t_d(\vec{r})} \frac{1}{\dot{G}_R(t_1,f,t_d(\vec{r}'))} = -\frac{j2\pi N}{\dot{G}_R^2(t_1,f,t_d(\vec{r}'))} \int_{-\infty}^{\infty} \Delta f \dot{G}(\vec{r}, \Delta f - f'') \dot{G}^*(\vec{r}, -f'') \sigma^0(\Delta f, \vec{r}) \times \qquad (27)$$
$$\times \dot{\Pi}(\Delta f - f'') \dot{\Pi}^*(-f'') \int_{-\infty}^{\infty} \dot{P}(\Delta f - f'' - f) \dot{P}^*(-f'' - f) df e^{-j2\pi \Delta f(t_1 - t_d(\vec{r}))} d\Delta f.$$

Substitute Equations (26) and (27) into Equation (25) and obtain the likelihood equation in the frequency–time domain

$$\int_0^T \int_{-\infty}^{\infty} \frac{\begin{bmatrix} \int_{-\infty}^{\infty} \Delta f \dot{G}(\vec{r}, \Delta f - f) \dot{G}^*(\vec{r}, -f) \\ \times \sigma^0(\Delta f, \vec{r}) \dot{\Pi}(\Delta f - f) \dot{\Pi}^*(-f) \\ \times \int_{-\infty}^{\infty} \dot{P}(\Delta f - f - f') \\ \times \dot{P}^*(-f - f') df' e^{-j2\pi \Delta f(t_1 - t_d(\vec{r}))} d\Delta f \end{bmatrix}}{\dot{G}_R(t_1,f,t_d(\vec{r}'))} df dt_1 = \int_0^T u(t_1) \int_{-\infty}^{\infty} \frac{1}{\dot{G}_R^2(t_1,f,t_d(\vec{r}'))} \int_{-\infty}^{\infty} \Delta f \dot{G}(\vec{r}, \Delta f - f) \times \qquad (28)$$
$$\times \dot{G}^*(\vec{r}, -f) \sigma^0(\Delta f, \vec{r}) \dot{\Pi}(\Delta f - f) \dot{\Pi}^*(-f) \int_{-\infty}^{\infty} \dot{P}(\Delta f - f - f') \dot{P}^*(-f - f') df' \times \dot{P}^*(-f'' - f) df e^{-j2\pi \Delta f(t_1 - t_d(\vec{r}))} d\Delta f.$$

For further calculations, we introduce the following notation:

$$\dot{Q}(\vec{r}, -f, t_1 - t_d(\vec{r})) = \int_{-\infty}^{\infty} \Delta f \dot{G}(\vec{r}, \Delta f - f) \sigma^0(\Delta f, \vec{r}) \dot{\Pi}(\Delta f - f) \int_{-\infty}^{\infty} \dot{P}(\Delta f - f - f') \dot{P}^*(-f - f') df' e^{-j2\pi \Delta f(t_1 - t_d(\vec{r}))} d\Delta f,$$

$$Q(\vec{r}, t_d(\vec{r})) = \int_0^T \int_{-\infty}^{\infty} \frac{\begin{bmatrix} \dot{G}^*(\vec{r}, -f) \dot{\Pi}^*(-f) \\ \times \dot{Q}(\vec{r}, -f, t_1 - t_d(\vec{r})) \end{bmatrix}}{\dot{G}_R(t_1,f,t_d(\vec{r}'))} df dt_1.$$

The function $Q(\vec{r}, t_d(\vec{r}))$ contains information about the delay time from each elementary section (within the area irradiated by the radiation pattern) of the underlying surface. Usually, for the practical use of altimeters, it is necessary to have a delay time from the nearest point, which can be obtained through integration over the irradiation area, that is, to change $Q(\vec{r}, t_d(\vec{r}))$ to $\int_D Q(\vec{r}, t_d(\vec{r})) d\vec{r} = Q(t_d)$. Then Equation (28) can be rewritten as follows

$$Q(t_d) = \int_D \int_0^T u(t_1) \int_{-\infty}^{\infty} \frac{\dot{\Pi}^*(-f) \dot{G}^*(\vec{r}, -f)}{\dot{G}_R^2(t_1,f,t_d(\vec{r}'))} \int_{-\infty}^{\infty} \Delta f \dot{G}(\vec{r}, \Delta f - f) \sigma^0(\Delta f, \vec{r}) \dot{\Pi}(\Delta f - f) \times \qquad (29)$$
$$\times \int_{-\infty}^{\infty} \dot{P}(\Delta f - f - f') \dot{P}^*(-f - f') df' \times e^{-j2\pi \Delta f(t_1 - t_d(\vec{r}))} d\Delta f \dot{U}^*(j2\pi f) df dt_1 d\vec{r}.$$

For further physical interpretation of the signal processing algorithm in the right part of likelihood Equation (29), we introduce some physically based assumptions:

(1) The amplitude–phase current distribution in the antenna aperture is uniform. Then the radiation pattern of the antenna in the coordinates of the underlying surface can be represented by an expression

$$G(x', f) = G(x'(\vartheta_x), f) = G\left(h \frac{\cos \alpha}{\cos \gamma}, f\right) = \frac{fX}{c} \text{sinc}\left(\pi h \frac{\cos \alpha}{\cos \gamma} \frac{fX}{c}\right) \qquad (30)$$

where fx/c is an analogue of spatial frequencies, which in ultra-wideband radar depends on both spatial coordinates and frequency; $x \in [-X/2, X/2]$; X is the size of the antenna along the axis 0x. It should be noted that according to Equation (30), the radiation pattern becomes more directional as the frequency increases, because the size of the antenna in wavelengths increases;

(2) The function $\sigma^0\left(\Delta f, \vec{r}\right)$ for most real surfaces is described by polynomials, but for the interpretation of the algorithm it is sufficient to consider it as a constant, i.e., $\sigma^0\left(\Delta f, \vec{r}\right) = 1$;

(3) The range of operating frequencies is limited by the function $\dot{\Pi}(f)$, which can be considered uniformly passable in the frequency range from F_{min} to F_{max}:

$$\Pi(f) = \begin{cases} 1 & F_{min} \leq |f| \leq F_{max}; \\ 0 & |f| < F_{min} \,\&\, |f| > F_{max}. \end{cases}$$

(4) Let the radio pulse envelop be uniform, then

$$\int_{-\infty}^{\infty} P(\Delta f - f - f')P(f + f')df' = T^2 \mathrm{sinc}(\pi f T)$$

Taking into account the introduced assumptions, the inner part of the integral (29), which can be represented in the following form

$$\frac{f X^2 T^2}{c^2} \mathrm{sinc}\left(\pi h \frac{\cos\alpha}{\cos\gamma} \cdot \frac{-fX}{c}\right) \dot{\Pi}^*(-f) e^{-j2\pi f(t_1 - t_d(\vec{r}))} \dot{M}\left(t_1 - t_d\left(\vec{r}\right)\right),$$

is a time–frequency–space function that describes the part of the decorrelating filter, where

$$\dot{M}\left(t_1 - t_d\left(\vec{r}\right)\right) = \int_{-\infty}^{\infty} \Delta f(\Delta f - f) \times \mathrm{sinc}\left(\pi h \frac{\cos\alpha}{\cos\gamma} \cdot \frac{(\Delta f - f)X}{c}\right) \dot{\Pi}(\Delta f - f) \mathrm{sinc}(\pi(\Delta f - f)T) e^{-j2\pi(\Delta f - f)(t_1 - t_d(\vec{r}))} d\Delta f.$$

Then likelihood Equation (29) should be represented as follows:

$$Q(t_d) = \int_D \int_0^T u(t_1) \int_{-\infty}^{\infty} \dot{Z}\left(f, \vec{r}, t_1 - t_d\left(\vec{r}\right)\right) \dot{U}^*(j2\pi f) e^{-j2\pi f(t_1 - t_d(\vec{r}))} df dt_1 d\vec{r} = \int_D \int_0^T u(t_1) u_Z\left(t_1 - t_d\left(\vec{r}\right)\right) dt_1 d\vec{r}, \quad (31)$$

where $u_Z\left(t_1 - t_d\left(\vec{r}\right)\right)$ is the observation after decorrelation in a filter with an amplitude–frequency response

$$\dot{Z}\left(f, \vec{r}, t_1 - t_d\left(\vec{r}\right)\right) = \frac{\dot{\Pi}^*(-f) \dot{G}^*\left(\vec{r}, -f\right)}{\hat{G}_R^2\left(t_1, f, t_d\left(\vec{r}'\right)\right)} \int_{-\infty}^{\infty} \Delta f \dot{G}\left(\vec{r}, \Delta f - f\right) \sigma^0\left(\Delta f, \vec{r}\right) \times$$
$$\times \dot{\Pi}(\Delta f - f) \int_{-\infty}^{\infty} \dot{P}(\Delta f - f - f') \dot{P}^*(-f - f') df' \, e^{-j2\pi(\Delta f - f)(t_1 - t_d(\vec{r}))} d\Delta f.$$
(32)

The right part of Equation (31) contains a signal processing algorithm to determine a function that depends on the delay time. However, considering the peculiarities of the ultra-wideband signals used in the problem being solved, it is necessary to further consider this algorithm, taking into account the following considerations. The internal correlation integral $\int_0^T u(t_1) u_Z\left(t_1 - t_d\left(\vec{r}\right)\right) dt_1$ is always close to zero, except for the case when the condition $t_1 - t_d\left(\vec{r}\right) = t_1$ is fulfilled. In addition, we neglect the decorrelation operation and move from $u_Z\left(t_1 - t_d\left(\vec{r}\right)\right)$ to $u\left(t_1 - t_d\left(\vec{r}\right)\right)$. Then Equation (31) can be represented as follows:

$$Q(t_d) = \int_D \int_0^T u^2\left(t_1 - t_d\left(\vec{r}\right)\right) dt_1 d\vec{r} = \begin{vmatrix} t_1 - t_d\left(\vec{r}\right) = z \\ dt_1 = dz \\ t_1 = 0 \quad z = -t_d\left(\vec{r}\right) \\ t_1 = T \quad z = T - t_d\left(\vec{r}\right) \end{vmatrix} = \int_D d\vec{r} \int_{-t_d(\vec{r})}^{T - t_d(\vec{r})} u^2(z) dz. \quad (33)$$

Algorithm (33) actually involves the calculation of a parameter proportional to the signal energy, and the averaging of the received energies from all areas of the underlying surface. However, here there is an uncertainty about the need to immediately calculate the detection threshold of the reflected signal with energy calculation in order to distinguish it from the background of the receiver's noise energy. In practice, this threshold can be chosen based on heuristic considerations.

3.2. Simulation Results

The structural diagram of the radar used for simulation is shown in Figure 2. The work of the model is as follows. The Envelop Pulse block generates an envelope of a noise pulse signal, which is then sent to the Signal Noise Generator block, which forms a wideband pulse stochastic signal. The generated radio pulse passes through the Signal propagation medium unit. In this unit, the signal is delayed for a time equivalent to a distance to the underlying surface of 1476 feet. This block also takes into account the dissipative properties of the signal propagation medium, as well as the fact that the radio pulse is reflected from the extended surface. The delayed signal is made noisy by the Internal Noise block and goes to the Signal Processing block, which performs the transformation in accordance with the algorithm (33). As mentioned earlier, algorithm (33) does not allow direct estimation of the range or the corresponding signal delay time. Thus, an envelope is observed at the output of the Signal processing block, which is further detected and based on which the delay time and the corresponding altitude to the underlying surface are calculated.

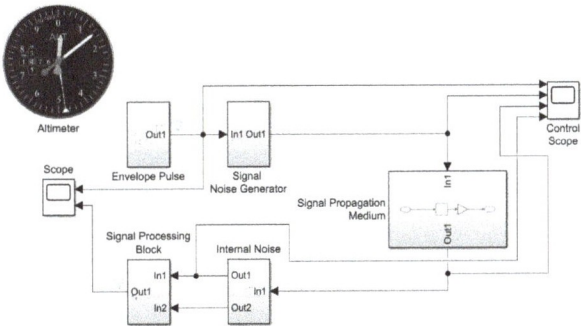

Figure 2. A simulation model of a wideband altimeter containing a transceiver part and realizing signal distortion during its propagation in the environment and reflection from the underlying surface.

The diagrams of the circuit operation, which describe the previously mentioned sequence, are shown in Figure 3, and the result of the circuit operation is shown in Figure 4.

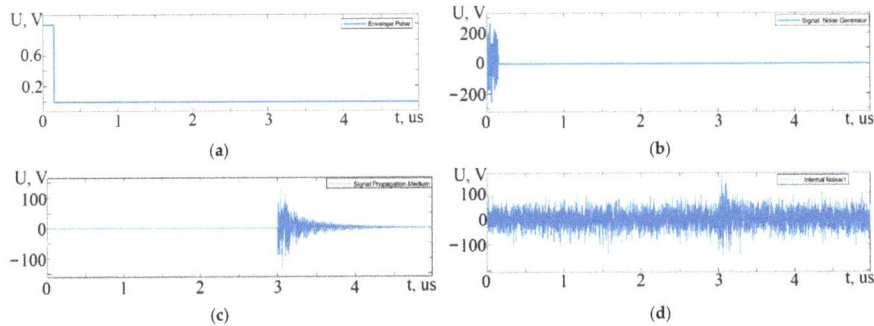

Figure 3. Signals at the blocks output: (**a**) pulse envelop; (**b**) signal noise generator; (**c**) signal propagation medium; and (**d**) internal noise.

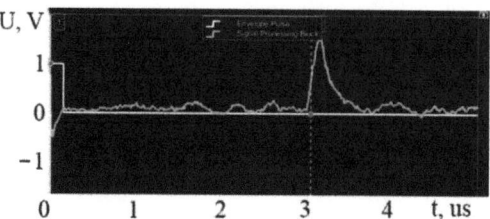

Figure 4. The time relationship between the emitted radio pulse envelope (yellow line) and the received reflected radio pulse envelope (blue line), the delay between which allows you to calculate the altitude to the underlying surface.

According to the obtained results in Figures 3 and 4, it follows that considering the dissipative properties of the atmosphere and reflection from an extended source leads to the spreading of the pulse in time (Figure 3c). The envelope of the signal at the radar output has also changed its shape and its falling front is more like an exponential function.

Similar changes will be observed in the case when there is a significant difference in height within the area that reflects the signal (for example, the "forest-grassland" border). However, this fundamentally does not change the result and the height measurement will take place to the nearest front of the reflector.

The conducted modeling allows us to state that the synthesized signal processing algorithm and the scheme implemented in accordance with it fully meet the formulated task. It should also be noted that the altimeter determines the altitude according to the value specified during modeling with a certain error.

4. Discussion

One of the promising paths in the development of many types of modern radar systems is the transition from narrow-band deterministic to wide- and ultra-wideband stochastic probing signals. In airborne radio altimeters, such a transition can increase the accuracy of determining the required altitude parameter and significantly improve the overall noise immunity of the complex. However, such changes require a more detailed study and search for optimal and quasi-optimal signal processing algorithms, which is the subject of this article.

The calculation of the optimal signal processing algorithm for the mentioned system is performed using the maximum likelihood method. To obtain reliable results, this method requires the most accurate determination of the initial data, namely, the models of useful signals and noises used, their statistical characteristics and general physically reasoned assumptions. All initial data and preliminary calculations are presented in the second section of the article.

It should be noted that the transition from deterministic probing signals to stochastic ones has significantly complicated the overall solution of the optimization problem. Therefore, the desired delay time parameter (or advisory altitude) is usually present explicitly in the likelihood functional for the case of deterministic signals. However, for the current case of a stochastic probing signal, the formalization of the delay time is significantly complicated by the impossibility of representing the reference signal in the form of a model or an analytical record. As a result, in the likelihood Equation (28) there is no explicit delay time $t_d(\vec{r})$, and the altitude calculation is possible using indirect parameters that depend on it. Thus, the resulting optimal algorithm (31) and quasi-optimal algorithm (33) provide for the delay time determination based on the received signal envelope $Q(t_d)$ calculation. The obtained result is well consistent with practice since many modern pulsed radar systems are also based on detecting the envelope of the received signal and further calculating the time between the signal emitted and received.

The main difference between the obtained optimal (31) and quasi-optimal (33) algorithms is the decorrelation operation of the received signal in a decorrelating filter with frequency response (32). The decoreleation operation in the algorithm (31) improves the potential signal envelope $Q(t_d)$ calculating accuracy, but its implementation in practice is extremely difficult due to the high complexity of obtaining all the decorrelating filter parameters in real time, for example, the current radar cross-section $\sigma^0\left(\Delta f, \vec{r}\right)$ of the underlying surface. Therefore, in practice, it is advisable to implement algorithm (33), which is devoid of decorating filtering and requires only averaging the energies of the signals received from all sections of the underlying surface. This greatly simplifies the technical implementation at the cost of negligible loss in accuracy. However, determining the degree of degradation in accuracy requires additional research and is not the goal of this work.

The overall performance of the quasi-optimal algorithm (33) is verified by simulation using the simulation model shown in Figure 2. The general principle of the flight altitude determination algorithm is to calculate the delay time between the envelope of the emitted stochastic radio pulse and the envelope of the received one, which for the simulation case are shown in Figure 4. Therefore, the delay between the two envelopes is approximately 3×10^{-6} s, that can be converted into an altitude in feet using the well-known formula

$$H = 3.28084 \frac{ct_d}{2}.$$

In the simulation, the calculated altitude is 1476 feet, that is displayed on the altimeter block in Figure 2. From the obtained simulation results, we can conclude that the signal-processing algorithm (33) is operable, however, it requires the introduction of an additional operation to determine the delay time between the envelopes of the emitted and received radio pulses, and also assumes that the detection threshold of the received signal is known. These features can be attributed to the shortcomings of the Equation (33) implementation; however, in the practical implementation of radars, it is not difficult to implement a block for accurately determining the delay time between envelopes. The threshold for detecting the envelope of the received signal is also often determined heuristically. Thus, the implementation of an altimeter with the signal processing algorithm (33) is possible, and the accuracy of the altitude determining will largely depend on the selected received signal detecting threshold and on the accuracy of the block for determining the delay time.

It should also be noted that the obtained structure of the radar, with the exception of the use of broadband stochastic probing signals and the corresponding high-frequency paths, largely corresponds both to the theory of other authors [14], and to the existing serial samples of airborne altimeters [29,30].

Future research development. As part of further work, calculations of the potential accuracy of the obtained algorithm will be performed. Also, taking into account that the use of serial samples of radio electronic devices is quite relevant today in the technical implementation of receivers, the task of synthesizing the altitude estimation algorithm for a given system with a partially given structure of the input path will be considered. Currently, an experimental model of a radio altimeter operating in the wavelength range of 3 mm is being developed, which will allow to examine the general possibility of practical implementation of the obtained algorithms, as well as to determine the peculiarities of their implementation in final systems.

5. Conclusions

In this paper, for the first time, an algorithm for processing wideband pulsed stochastic signals in a radio altimeter is synthesized using the maximum likelihood method. The use of wideband signals in the given system led to some difficulties in solving the problem. Firstly, it concerns the difficulty of formalizing the delay time parameter in the likelihood functional for a noisy reference signal, since it could not be represented in the form of a model or an analytical equation. At the same time, due to the pulse signal being nonstationarity, it became necessary to calculate the corresponding inversion equation, that is performed for the first time. However, regardless of the formal difficulties in the approach to solving the problem, the synthesized signal processing algorithm is quite clear both for analysis and for its further technical implementation.

In accordance with the obtained algorithm, a simulation model of the radar is developed, and its simulation is carried out considering the unevenness of the underlying surface and the dispersed properties of the signal propagation medium. It is worth noting that the resulting algorithm does not allow direct estimation of the delay time parameter or the corresponding distance to the underlying surface, but only evaluates the envelope of the received signal. Therefore, its practical implementation should provide additional algorithms for estimating the delay time to the detected envelope. However, the resulting algorithm is quite convenient and understandable for technical implementation on a modern radio element base.

Author Contributions: Conceptualization, V.V., V.P. and O.O.; methodology, V.V. and S.Z.; software, E.T. and A.P.; validation, O.U., A.P. and A.H.; formal analysis, V.P. and S.Z.; investigation, O.O. and A.P.; resources, E.T.; data curation, A.H.; writing—original draft preparation, V.P. and O.O.; writing—review and editing, S.Z. and O.U.; visualization, E.T.; supervision, V.V.; project administration, V.P.; funding acquisition, O.U. All authors have read and agreed to the published version of the manuscript.

Funding: The work was funded by the Ministry of Education and Science of Ukraine, the state registration numbers of the projects are 0122U200469 and 0121U109598.

Institutional Review Board Statement: Not applicable.

Informed Consent Statement: Not applicable.

Data Availability Statement: Not applicable.

Conflicts of Interest: The authors declare no conflict of interest.

References

1. Chen, M.; Zhang, Y.; Chen, Y. Review on civil aviation safety investment research. In Proceedings of the 2016 11th International Conference on Reliability, Maintainability and Safety (ICRMS), Hangzhou, China, 26–28 October 2016; pp. 1–5.
2. Lungu, M.; Lungu, R. Complete landing autopilot having control laws based on neural networks and dynamic inversion. In Proceedings of the 2017 18th International Carpathian Control Conference (ICCC), Sinaia, Romania, 28–31 May 2017; pp. 17–22.
3. Watkins, D.; Gallardo, G.; Chau, S. Pilot Support System: A Machine Learning Approach. In Proceedings of the 2018 IEEE 12th International Conference on Semantic Computing (ICSC), Laguna Hills, CA, USA, 31 January–2 February 2018; pp. 325–328.
4. Bolanakis, D.E. MEMS barometers and barometric altimeters in industrial, medical, aerospace, and consumer applications. *IEEE Instrum. Meas. Mag.* **2017**, *20*, 30–55. [CrossRef]
5. Kai, L.; Jun-Jie, L.; Jing, W.; Xiao-Jun, W. Research on Augmented Reality Technology of Heli-copter Aided Navigation Based on Lidar. In Proceedings of the 2021 IEEE 7th International Conference on Virtual Reality (ICVR), Foshan, China, 20–22 May 2021; pp. 373–379.
6. Zhao, J.; Li, Y.; Hu, D.; Pei, Z. Design on altitude control system of quad rotor based on laser radar. In Proceedings of the 2016 IEEE International Conference on Aircraft Utility Systems (AUS), Beijing, China, 8–14 October 2016; pp. 105–109.
7. Brockers, R. Autonomous Safe Landing Site Detection for a Future Mars Science Helicopter. In Proceedings of the 2021 IEEE Aerospace Conference (50100), Big Sky, MT, USA, 6–13 March 2021; pp. 1–8.
8. Rahman, A.M.; Hossain, S.; Tuku, I.J.; Hossam-E-Haider, M.; Amin, M.S. Feasibility study of GSM network for tracking low altitude helicopter. In Proceedings of the 2016 3rd International Conference on Electrical Engineering and Information Communication Technology (ICEEICT), Dhaka, Bangladesh, 22–24 September 2016; pp. 1–8.
9. Videmsek, A.; de Haag, M.U.; Bleakley, T. Evaluation of RADAR altimeter-aided GPS for precision approach using flight test data r. In Proceedings of the Digital Avionics Systems Conference (DASC), San Diego, CA, USA, 8–12 September 2019; pp. 1–10.
10. Liu, B.; Meng, Z.; Zhou, Y. Nonlinear path following for unmanned helicopter. In Proceedings of the IEEE International Conference on Information and Automation, Macau, China, 18–20 July 2017; pp. 737–743.
11. Chen, Y.; Wang, C.; Zeng, W.; Wu, Y. Horizontal Nonlinear Path Following Guidance Law for a Small UAV with Parameter Optimized by NMPC. *IEEE Access* **2021**, *9*, 127102–127116. [CrossRef]
12. Ragi, S.; Mittelmann, H.D.; Zhou, Y. Mixed-integer nonlinear programming formulation of a UAV path optimization problem. In Proceedings of the 2017 American Control Conference (ACC), Seattle, WA, USA, 24–26 May 2017; pp. 406–411.

13. *686-2017—IEEE Standard for Radar Definitions*; IEEE: New York, NY, USA, 2017; pp. 1–54.
14. Madhupriya, G.; Lavanya, K.S.; Vennisa, V.; Raajan, N.R. Implementation of Compressed Wave Pulsed Radar Altimeter in Signal Processing. In Proceedings of the 2019 International Conference on Computer Communication and Informatics (ICCCI), Coimbatore, India, 23–25 January 2019; pp. 1–5.
15. Li, Y.; Hoogeboom, P.L.; Dekker, P.L.; Mok, S.-H.; Guo, J.; Buck, C. CubeSat Altimeter Constellation Systems: Performance Analysis and Methodology. *IEEE Trans. Geosc. Rem. Sens.* **2022**, *60*, 1–19. [CrossRef]
16. Mogyla, A.; Kantsedal, V. Estimation of the Parameters of the Stochastic Probing Radio Signal Reflected by the Target. In Proceedings of the 2020 IEEE Ukrainian Microwave Week (UkrMW), Kharkiv, Ukraine, 22–27 June 2020; pp. 379–383.
17. Ravenscroft, B.; Blunt, S.D.; Allen, C.; Martone, A.; Sherbondy, K. Analysis of spectral notching in FM noise radar using measured interference. In Proceedings of the International Conference on Radar Systems (Radar 2017), Belfast, UK, 23–26 October 2017; pp. 1–6.
18. Wasserzier, C.; Wojaczek, P.; Cristallini, D.; Worms, J.; O'Hagan, D. Doppler-Spread Clutter Suppression in Single-Channel Noise Radar. In Proceedings of the 2019 International Radar Conference (RADAR), Toulon, France, 23–27 September 2019; pp. 1–4.
19. Zhang, L.; Kuang, Q.; Chang, M.; Zhang, D. A 10-GS/s Sample and Hold System for High-Speed Time-Interleaved ADC. In Proceedings of the 2018 Eighth International Conference on Instrumenta-tion & Measurement, Computer, Communication and Control (IMCCC), Harbin, China, 19–21 July 2018; pp. 705–708.
20. Genschow, D.; Kloas, J. Evaluation of a UWB radar interface for low power radar sensors. In Proceedings of the 2015 European Radar Conference (EuRAD), Paris, France, 9–11 September 2015; pp. 321–324.
21. Wang, D. A Gm-Compensated 46–101 GHz Broadband Power Amplifier for High-Resolution FMCW Radars. In Proceedings of the 2021 IEEE International Symposium on Circuits and Systems (ISCAS), Daego, Korea, 22–28 May 2021; pp. 1–5.
22. Rachid, E.A.; Farha, R.M. Ultra-wideband: Very simple printed antennas with Windows in K, Ka and Q bands. In Proceedings of the 2016 IEEE Conference on Antenna Measurements & Applications (CAMA), New York, NY, USA, 23–27 October 2016; pp. 1–4.
23. Tarchi, D.; Lukin, K.; Fortuny-Guasch, J.; Mogyla, A.; Vyplavin, P.; Sieber, A. SAR Imaging with Noise Radar. *IEEE Trans. Aerosp. Electron. Syst.* **2010**, *46*, 1214–1225. [CrossRef]
24. Ostroumov, I.; Kuzmenko, N.; Sushchenko, O.; Pavlikov, V.; Zhyla, S.; Solomentsev, O.; Zaliskyi, M.; Averyanova, Y.; Tserne, E.; Popov, A.; et al. Modelling and simulation of DME navigation global service volume. *Adv. Space Res.* **2021**, *68*, 3495–3507. [CrossRef]
25. Pavlikov, V.; Belousov, K.; Zhyla, S.; Tserne, E.; Shmatko, O.; Sobkolov, A.; Vlasenko, D.; Kosharskyi, V.; Odokiienko, O.; Ruzhentsev, M. Radar imaging complex with sar and asr for aerospace vechicle. *Radioel. Comp. Syst.* **2021**, *3*, 63–78. [CrossRef]
26. Pavlikov, V.; Volosyuk, V.; Zhyla, S.; Tserne, E.; Shmatko, O.; Sobkolov, A. Active-Passive Radar for Radar Imaging from Aero-space Carriers. In Proceedings of the 2016 IEEE EUROCON 2021—19th International Conference on Smart Technologies, Lviv, Ukraine, 6–8 July 2021; pp. 18–24.
27. Zhyla, S.; Volosyuk, V.; Pavlikov, V.; Ruzhentsev, N.; Tserne, E.; Popov, A.; Shmatko, O.; Havrylenko, O.; Kuzmenko, N.; Dergachov, K.; et al. Statistical synthesis of aerospace radars structure with optimal spatio-temporal signal processing, extended observation area and high spatial resolution. *Radioel. Comp. Syst.* **2022**, *1*, 178–194. [CrossRef]
28. Volosyuk, V.; Kravchenko, V. *Theory of Radio-Engineering Systems of Remote Sensing and Radar*; Fizmatlit: Moscow, Russia, 2008.
29. Compact Radar Altimeter. Available online: https://electronics.leonardo.com/documents/16277707/18391547/Compact+Radar+Altimeter+%28MM09113%29_LQ.pdf (accessed on 18 March 2022).
30. Radar Altimeter. Available online: https://www.freeflightsystems.com/blog/product/radar-altimeters/ (accessed on 25 March 2022).
31. Arslan, S.; Yıldırım, B.S. A Broadband Microwave Noise Generator Using Zener Diodes and a New Technique for Generating White Noise. *IEEE Microw. Wirel. Compon. Lett.* **2018**, *28*, 329–331. [CrossRef]
32. Pritsker, D.; Cheung, C.; Neoh, H.S.; Nash, G. Wideband Programmable Gaussian Noise Generator on FPGA. In Proceedings of the 2019 IEEE National Aerospace and Electronics Conference (NAECON), Dayton, OH, USA, 15–19 July 2019; pp. 412–415.
33. Kravchenko, V.; Kutuza, B.; Volosyuk, V.; Pavlikov, V.; Zhyla, S. Super-resolution SAR imaging: Optimal algorithm synthesis and simulation results. In Proceedings of the 2017 Progress in Electromagnetics Research Symposium—Spring (PIERS), St. Petersburg, Russia, 22–25 May 2017; pp. 419–425.
34. Pavlikov, V.; Zhyla, S.; Odokienko, O. Structural optimization of Dicke-type radiometer. In Proceedings of the 2016 II International Young Scientists Forum on Ap-plied Physics and Engineering (YSF), Kharkiv, Ukraine, 10–14 October 2016; pp. 171–174.
35. Volosyuk, V.; Zhyla, S.; Pavlikov, V.; Ruzhentsev, N.; Tserne, E.; Popov, A.; Shmatko, O.; Dergachov, K.; Havrylenko, O.; Ostroumov, I.; et al. Optimal Method for Polarization Selection of Stationary Objects against the Background of the Earth's Surface. *Int. J. Electron. Telecommun.* **2022**, *68*, 83–89.
36. Tsopa, A.I.; Ivanov, V.K.; Leonidov, V.I.; Maleshenko, Y.I.; Pavlikov, V.V.; Ruzhentsev, N.V.; Zarudniy, A.A. The research program of millimetric radio waves attenuation characteristics on perspective communication lines of Ukraine. In Proceedings of the 2016 13th International Conference on Modern Problems of Radio Engineering, Telecommunications and Computer Science (TCSET), Lviv, Ukraine, 23–26 February 2016; pp. 638–642.

37. Pavlikov, V.V.; Volosyuk, V.K.; Zhyla, S.S.; Van, H.N. Active Aperture Synthesis Radar for High Spatial Resolution Imaging. In Proceedings of the 2018 9th International Conference on Ultrawideband and Ultrashort Impulse Signals (UWBUSIS), Odessa, Ukraine, 4–7 September 2018; pp. 252–255.
38. Astola, J.T.; Egiazarian, K.O.; Khlopov, G.I.; Khomenko, S.I.; Kurbatov, I.V.; Morozov, V.Y.; Totsky, A.V. Application of Bispectrum Estimation for Time-Frequency Analysis of Ground Surveillance Doppler radar Echo Signals. *IEEE Trans. Instrum. Meas.* **2008**, *57*, 1949–1957. [CrossRef]

Article

Capacitive Water-Cut Meter with Robust Near-Linear Transfer Function

Oleksandr Zabolotnyi *, Vitalii Zabolotnyi and Nikolay Koshevoy

Department of Intellectual Instrumentation Systems and Quality Management, National Aerospace University «Kharkiv Aviation Institute», 61070 Kharkiv, Ukraine; v.zabolotnyi@khai.edu (V.Z.); m.koshovyi@khai.edu (N.K.)
* Correspondence: pretorian14@ukr.net

Abstract: The water content in fuel–water emulsions can vary from 10% to 30%, and is under control during the process of emulsification. The main task of this study was to obtain near-linear static function for a water-cut meter with capacitive sensors, and to provide it with effective type-uncertainty compensation during the process of water–fuel emulsion moisture control. To fulfill the capacitive measurements, two capacitive sensors in the measuring channel and two capacitive sensors in the reference channel were used. The method of least squares and general linear regression instruments were used to obtain robust and near-linear transfer function of the capacitive water-cut meter. The prototype product of the water-cut meter was developed with the purpose of fulfilling multiple moisture measurements and checking the workability of the new transfer function. Values of moisture content for the new transfer function and the closest analog were compared with the help of dispersion analysis. The new transfer function provided minimal dispersions of repeatability and adequacy, and minimal *F*-test values, proving its better capability for type-uncertainty compensation and better adequacy for the nominal linear transfer function of the water-cut meter.

Keywords: fuel–water emulsion; capacitive water-cut meter; near-linear transfer function; general linear regression; robust properties; dispersion analysis

1. Introduction

Emulsified fuels have a clear effect on the combustion process, and are one of the modern combustible systems that offer prospects for new technologies. This has been exploited recently in diesel engines, which have been used for a long time as the main source of driving power due to their reliable design and high fuel-saving economy [1,2]. Significant results in the toxicity reduction of diesel engines' exhaust fumes were achieved by the application of fuels with water and alcohol additives [3]. Hence, emulsified fuels are mixtures of a combustible liquid, either oil or fuel as a continuous phase, and a smaller amount of water with or without a surfactant as a dispersed phase. Emulsion of petroleum fuel in water is the most common form of such emulsion [4]. In emulsified diesel fuel, the heat absorption by water vaporization causes a decrease in the local adiabatic flame temperature, and this reduces the chemical reaction in the gas phase to produce thermal NO_x [5]. Moreover, this helps to reduce the formation of soot, PM, CO, and HC [6,7]. Water–fuel emulsion is a system that includes water (dispersed phase with 0.1 ... 10 μm diameter of droplets) and diesel fuel as a dispersion medium [8–10], and the volumetric water content in the fuel–water emulsions can vary from 5% to 30%, and is an object of strict control during the process of emulsification [4,11–13].

Methods of moisture measurement can be divided into direct—based on dividing the substance under research into free water and dry material—and indirect, where moisture content can be defined based on the values of other measurements functionally connected with moisture [14–16]. To identify popular principles of moisture measurement, a detailed analysis of the world moisture meter market, produced in a recent period, was carried

out for bulk substances and oil products as objects of measurement. In total, 358 moisture meters for bulk materials and 64 moisture meters for oil products were detected. The dielcometer (capacitive) principle of moisture measurement remains forward-looking among all indirect methods [17].

A large number of factors influence moisture measurement accuracy. Among them, we traditionally emphasize the physicochemical composition of the combustible liquid (oil fuel as a continuous phase of the emulsion) [16,18,19]. It is not only the type of fuel that has an effect on its physicochemical composition, but the conditions of extraction and processing, etc., and obtaining a full list of factors for analytic forecasting is usually complicated.

A list of analytical methods has been developed to determine the moisture content of oils, based on the fact that water has appreciably different bulk physical characteristics than the oil matrix, e.g., density, dielectric permittivity, or refractive index. These methods are usually only suitable for the analysis of substances in which the composition of the oil matrix does not change significantly, but where the water-to-oil ratio of the matrix changes. For example, the water content of an oil-in-water emulsion can be determined by measuring the density or dielectric permittivity, because the density and dielectric permittivity of water are significantly higher than those of oil. If the composition of the oil matrix changes as well as the water content, then it may not be possible to accurately determine the moisture content of the oil, because more than one oil composition may give the same value for the physical property being measured. Unfortunately, it is impossible to obtain an analytic forecast of the chemical composition and different features of all of the emulsified fuels under research. This is why most indirect methods of moisture measurement have local effectiveness, and are not versatile.

At present, approximately 33% of moisture meters are represented by dielcometer measuring instruments with capacitive sensors. Moisture meters of this type have method error (type uncertainty), the value of which directly depends on the dielectric permittivity of the dry oil as a continuous phase of the emulsion under research (oil type, for example) [16,18]. The influence of type uncertainty on the result of moisture measurement is usually significant, and traditional methods of compensation can be effective when we know the composition of a substance. Such compensation, in general, can be fulfilled using a secondary measuring transducer by using complementary reference capacitors, dedicated analytic calculations, reference calibration curves stored in the secondary measuring transducers' memory, etc. It can all be used for the limited number of materials under research, and does not allow the moisture meter to be versatile for a wide range of fuel substances. In other cases, the effectiveness of traditional compensation methods essentially decreases, because of which further improvement of the existing methods of moisture measurement with a view to solving the problem of type-uncertainty compensation is a relevant and perspective mission [20]. In-depth analysis of modern moisture control approaches with the purpose of further development and improvement is still a relevant task, and the process of searching for new methods of water-cut sensing optimization is under constant continuation [21–23].

"Successful" modifications of capacitive moisture meters with capacitive sensors appeared years ago [24–26]. An idea to obtain capacitive measurements using two or more positions for one of the capacitor plates, with further direct comparison, seemed to be of special interest [27,28]. This approach helps to eliminate parasitic capacitances almost completely, compensate for leakage currents, and reduce the influence of fringe electric fields. The main task of this research is in achieving robust, near-linear transfer function for the comparison of methods of moisture measurement, as described in [17], and to provide effective type-uncertainty compensation during the process of water–fuel emulsion moisture control.

2. Materials and Methods

In capacitive sensors, where one of the capacitor plates should be placed in two or more positions [27], we first take an empty sensor, where a change in electric capacitance ΔC_0 should be defined before and after the distance between two plates is changed. Then, we define the change in electric capacitance ΔC_1 when a capacitive sensor is filled with a tested substance for the same positions of the capacitor plates. After that, the relation $\Delta C_1 / \Delta C_0$ should be calculated. In this method, only the accuracy of the capacitor plates' positioning would influence the uncertainty of the measurements. Thus, the transfer function of the moisture meter should be described with the following formula:

$$W = K \cdot \frac{C_3 - C_4}{C_2 - C_1}, \qquad (1)$$

where K is a normalizing coefficient [17] equal to $K = 28.599$ when we take the values of electric capacitances, equal to 15 pF for C_1 and C_4, and 50 pF for C_2 and C_3 when the capacitive sensors are empty.

The suggested design of the capacitive instrument measuring transducer is described below (Figure 1). As we can see, both of its sensors consist of a system of flat plates (1), where two pairs of flat plates belong to measuring capacitors C_1 and C_4, and the rest of the flat plates create another pair of measuring capacitors C_2 and C_3. All flat plates of equal length l are assembled inside two fluoroplastic rings (2) at an equal gap, designated as Z.

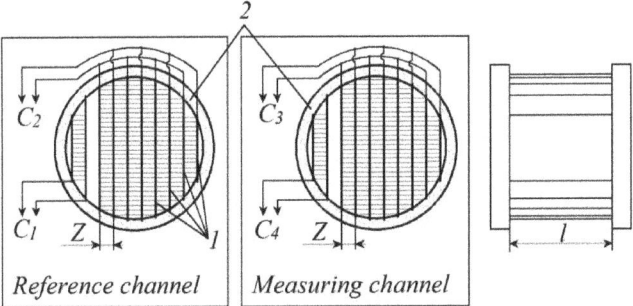

Figure 1. Capacitive instrument measuring transducer for moisture content (1—system of flat electrodes, 2—fluoroplastic rings).

The measuring channel should be filled with a probe of liquid fuel under moisture control, and the reference channel should be filled with the same substance, but previously dehydrated. Using the method mentioned above, it is necessary to measure the values of electric capacitances C_1, C_2, C_3, and C_4, and then to calculate the differences $C_2 - C_1$ and $C_3 - C_4$. When this is done, we can calculate the relationship of these differences $(C_3 - C_4)/(C_2 - C_1)$, and this would be an informative parameter for further processing, directly connected with moisture content.

To implement this idea in practice, moisture content measuring instruments should have a structure, as given in Figure 2. As we can see, four capacitive sensors of the measuring transducer are connected to the inputs of corresponding monostable multivibrators $MMV1 \ldots MMV4$, designed to convert values of electric capacitance into a duration of rectangular pulses. The microcontroller CPU controls the monostable multivibrators' work by generating rectangular pulses of a stable frequency on one of its output pins.

Each monostable multivibrator generates a sequence of rectangular pulses, the duration of which is in direct proportion with the value of electric capacitance, connected to its input (diagrams in Figure 3).

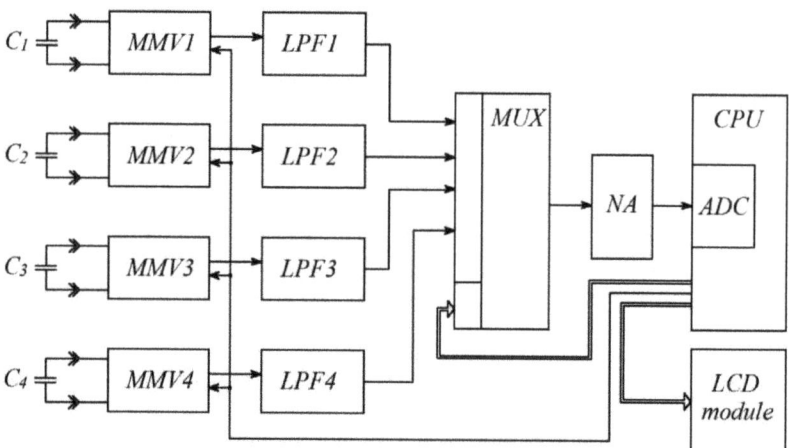

Figure 2. Structural electric circuit of the moisture meter (*MMV1* ... *MMV4*—monostable multivibrators, *LPF1* ... *LPF4*—low-pass filters, *MUX*—multiplexer, *NA*—normalizing amplifier, *ADC*—analog-to-digital converter).

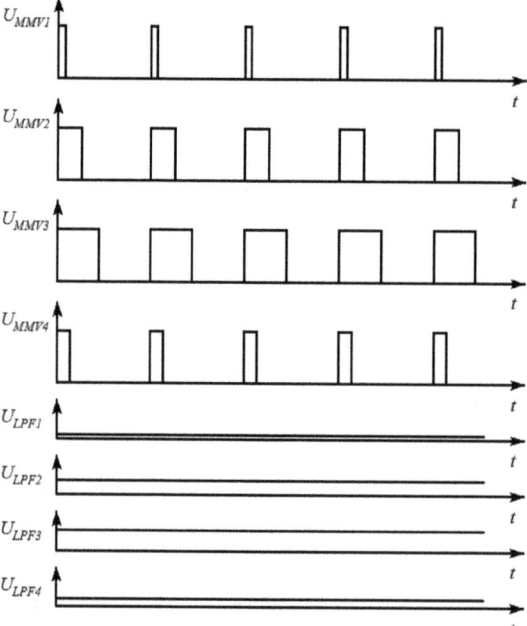

Figure 3. Operating diagrams for the analogue part of the moisture meter.

Low-pass filters convert pulse duration into DC voltage in proportion to the pulse duration amplitude. The multiplexer *MUX*, controlled by the *CPU*, commutes the outputs of the low-pass filters *LPF1* ... *LPF4* with the input of the normalizing amplifier *NA* which, in turn, provides levels of DC voltage compatible with the measuring range of the analog-to-digital converter *ADC* embedded in the *CPU*. The microcontroller *CPU*, after having four values of DC voltage proportional to the values of the four electric capacitances C_1, C_2, C_3, and C_4, calculates the moisture content of the substance under research, and displays the results on the LCD module's screen. To check the behavior of the transfer function (1) with different types of moist substances, we took four values of dielectric permittivity for imaginary dehydrated fuels ($\varepsilon_n = 2.0$; $\varepsilon_n = 2.5$; $\varepsilon_n = 3.0$; $\varepsilon_n = 3.5$) and four values of moisture content ($W = 0\%$; $W = 10\%$; $W = 20\%$; $W = 30\%$). The values of dielectric permittivity for moist substances, necessary to calculate the values of the four sensors' capacitances, were estimated with the help of the universal Wiener equation [29–32]. The values of capacitances C_1, C_2, C_3, and C_4, and the calculated values of moisture content defined with the help of the static function (1), are given in Table 1.

Table 1. Calculated values of sensors' capacitances and moisture content for the transfer function (1).

W, %	C_i, pF	ε_n			
		2.0	2.5	3.0	3.5
0	C_1	30	37.5	45	52.5
	C_2	100	125	150	175
	C_3	100	125	150	175
	C_4	30	37.5	45	52.5
	W_{calc}	0	0	0	0
		ε_n			
		2.614	3.252	3.885	4.512
10	C_1	30	37.5	45	52.5
	C_2	100	125	150	175
	C_3	130.7	162.6	194.25	225.6
	C_4	39.21	48.78	58.28	67.68
	W_{calc}	8.78	8.60	8.44	8.23
		ε_n			
		3.368	4.317	4.963	5.741
20	C_1	30	37.5	45	52.5
	C_2	100	125	150	175
	C_3	168.4	215.85	248.15	287.05
	C_4	50.52	64.76	74.45	86.12
	W_{calc}	19.56	20.78	18.71	18.31
		ε_n			
		4.317	5.324	6.305	7.262
30	C_1	30	37.5	45	52.5
	C_2	100	125	150	175
	C_3	215.85	266.2	315.25	363.1
	C_4	64.76	79.86	94.58	108.93
	W_{calc}	33.13	32.31	31.51	30.74

Graphs of received transfer functions for different moist substances, and their comparison with an ideal linear transfer function, are given in Figure 4.

As we can see, all transfer functions received with the help of Equation (1) are nonlinear, and the calculated values of moisture content are significantly different from nominal in all points, except for $W = 0\%$. Hence, at first, Equation (1) should be linearized.

To achieve near-linear dependence between moisture content W and the $(C_3 - C_4)/(C_2 - C_1)$ relationship, the method of least squares and instruments of general linear regression were used. The modified transfer function for the water-cut meter is given in Equation (2), where a and b are coefficients preliminarily defined in [33] ($a = 0.9622$, $b = 0.0371$):

$$W_m = -1.019 + 1.269 \cdot \frac{\left(\frac{C_3-C_4}{C_2-C_1} - a\right)}{b} - 0.00527 \cdot \frac{\left(\frac{C_3-C_4}{C_2-C_1} - a\right)^2}{b^2} - 0.000112 \cdot \frac{\left(\frac{C_3-C_4}{C_2-C_1} - a\right)^3}{b^3}. \quad (2)$$

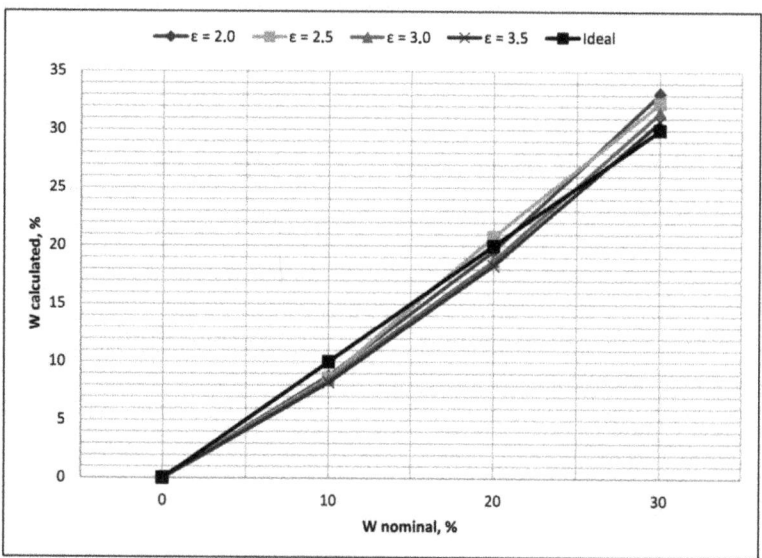

Figure 4. Calculated transfer functions, placed together with an ideal transfer function of the water-cut meter.

New values of moisture content, calculated using Formula (2), are given in Table 2.

Table 2. Calculated values of a static function (2).

W, %	W_m			
	$\varepsilon_n = 2.0$	$\varepsilon_n = 2.5$	$\varepsilon_n = 3.0$	$\varepsilon_n = 3.5$
0	0.268	0.268	0.268	0.268
10	10.230	10.045	9.859	9.674
20	20.850	20.470	20.085	19.723
30	30.658	30.211	29.476	29.280

The graphs of received transfer functions for different moist substances can be seen in Figure 5. If we compare Figures 4 and 5, it would be possible to say that the modified transfer function (2) is far more effective than the initial transfer function (1).

To prove this analytically, it seemed to be rational to calculate one of the goodness-of-fit parameters, e.g., root mean estimator (Equation (3)):

$$S = \sqrt{\frac{\sum(W_{inom} - W_{icalc})^2}{n}}. \tag{3}$$

For the data in Table 1, we can write:

$$\sum(W_{inom} - W_{icalc})^2 = (0-0)^2 \cdot 4 + (10-8.78)^2 + (10-8.60)^2 + (10-8.44)^2 +$$
$$+ (10-8.23)^2 + (20-19.56)^2 + (20-20.78)^2 + (20-18.71)^2 + (20-18.31)^2 +$$
$$+ (30-31.13)^2 + (30-32.31)^2 + (30-31.51)^2 + (30-30.74)^2 = 27.2978,$$

$$S = \sqrt{\frac{27.2978}{16}} = 1.3062 \%.$$

For the data in Table 2, the result would be $S = 0.4158\%$.

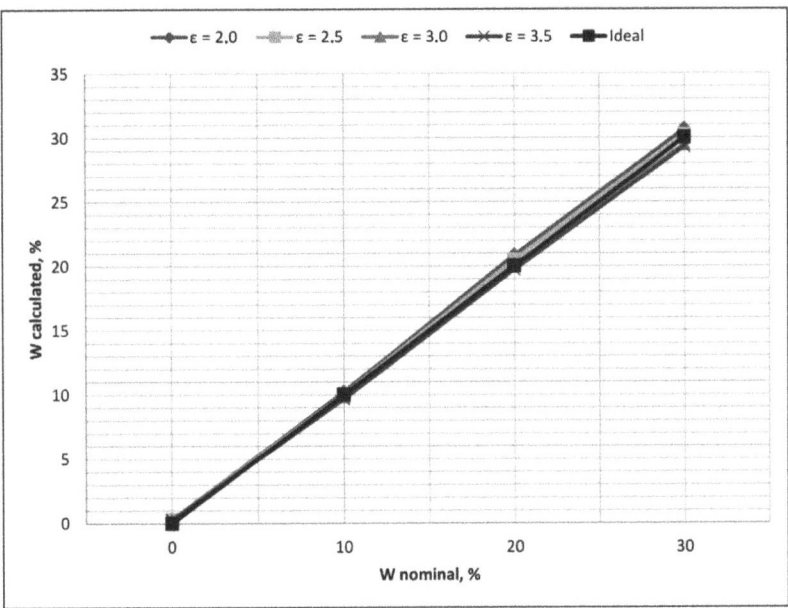

Figure 5. Modified transfer functions together with an ideal transfer function of the water-cut meter.

To carry out experimental research with the purpose of checking the workability of a transfer function (2), and to estimate the provided level of uncertainty in the moisture content measurements, it was necessary to obtain stable reference samples of oil products with different dielectric permittivity values ε_n in a dehydrated state, and values of moisture content equal to $W = 0\%$, $W = 10\%$, $W = 20\%$, and $W = 30\%$. We used diesel fuel with $\varepsilon_n = 2.01$ and mazut with $\varepsilon_n = 2.67$ for this purpose. Stable reference samples of diesel fuel and mazut with 10 μm diameter of water particles were obtained by adding appropriate volumes of water, with the help of a second-class precision pipette [34] and 30 min of mechanical mixing at $n = 3200$ rpm [35,36].

Sensors of a measuring transducer were assembled from four identical fluoroplastic rings with slots to insert flat stainless steel plates, which were glued to the internal surface of the appropriate pair of fluoroplastic rings with the help of superglue gel, and soldered with wires to create four capacitive sensors (Figure 6).

The process of moisture measurement was performed following the substitution method, where the value of the quantity being measured is not found directly from a reading of the measuring instrument, but rather from the magnitude of the standard, which is regulated in such a way that the reading of the measuring instrument remains the same when the quantity being measured is replaced by the standard. Commonly known, this method eliminates systematic errors and provides high accuracy, and the substitution method is extensively used in measuring electrical quantities, such as resistance, capacitance, and inductance. The connection scheme of the devices involved in the process of moisture measurement is given in Figure 7.

The setup consists of four capacitive sensors, designated as C_1, C_2, C_3, and C_4, one channeled capacitance into a DC-voltage transducer that performs a sequence of transformations "capacitance–pulse duration–dc voltage", a variable air capacitor (reference capacitor), an oscilloscope, a digital voltmeter, and an RLC meter.

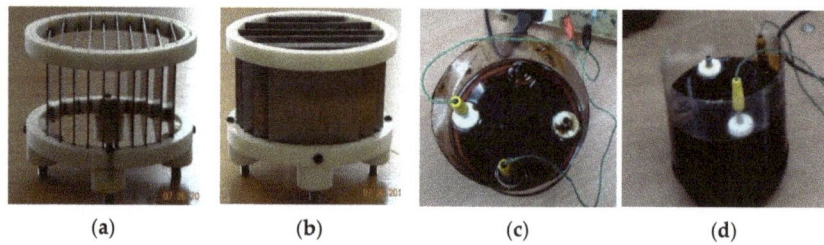

Figure 6. Prototype product of the capacitive measuring transducer: (**a**,**b**) in an empty state, (**c**) filled with diesel fuel (ε_n = 2.01), and (**d**) filled with mazut (ε_n = 2.67).

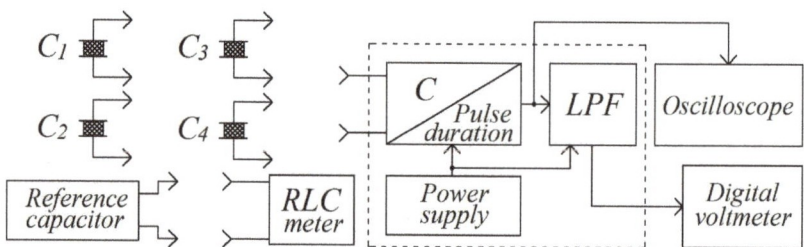

Figure 7. Connection of the devices involved in the moisture measurement: experimental setup.

From the very beginning, a part of the measuring transducer with two capacitive sensors C_1 and C_2 (reference channel) was filled with the original (with traces of water) sample of oil product (Figure 8a). Both capacitive sensors of the reference channel were connected one by one to the input of capacitance into the DC-voltage transducer, and appropriate DC voltage values were taken from the screen of a digital voltmeter and fixed by the operator. Then, the variable air capacitor (reference capacitor or standard) was connected instead of the capacitive sensors. Its capacitance was slowly changed in correspondence with the method of substitution until the moment when the DC voltage value on the voltmeter's screen became equal to the values detected previously (Figure 8b).

Figure 8. Process of moisture measurement: (**a**) connection of the first part of the measuring transducer (sensors C_1 and C_2 with dehydrated diesel fuel) to the input of a secondary capacitance into the DC-voltage transducer; (**b**) the reference capacitor's connection instead of sensors C_1 and C_2; (**c**) measuring the capacitance of a reference capacitor with the help of the RLC meter; (**d**) connection of the second part of the measuring transducer (sensors C_3 and C_4 with moist diesel fuel, W = 20%) to the input of a secondary capacitance into the DC-voltage transducer.

After that, the standard capacitor was disconnected from the transducer, and its capacitance was measured with the help of an accurate RLC meter at a 10 kHz frequency. The same process took place for the capacitive sensors in the measuring channel, filled with moist samples of oil products (Figure 8d). As a result, 10 measurements of electric capacitance were received from each capacitive sensor. The oscilloscope helped to control the correct operation of the capacitance into the pulse duration transducer. At first, this helped to choose the correct operating mode of the secondary transducer, when the duration of rectangular pulses was in direct proportion with the C_1, C_2, C_3, and C_4 capacitance values in the whole range of their variation. Furthermore, the presence of the oscilloscope was necessary to detect possible oscillation stops when measuring high moisture contents (i.e., over 20%).

3. Theory/Calculation

The next step was to check the ability of the new transfer function (2) to retain stability when working with experimental results of measurements with natural random variation. The values of electric capacitances taken from capacitive sensors were not the ideal deterministic values used in Table 1, but would always have some random variation. The presence of random variation (i.e., capacitance measurement uncertainty) influences the uncertainty of the calculated moisture content values; it can decrease the robustness of the transfer function (2) up to the moment when it becomes irrelevant [37–39]. To check the robustness of the transfer function (2) theoretically, we introduced symmetric variation into the values of electric capacitance. The range of variation was set at a level of 0.01% from the values of electric capacitance after the analysis of metrological performance for modern capacitance and RLC meters. The calculated values of electric capacitance for capacitive sensors placed in the reference channel are shown in Table 3, while those for the sensors of the measuring channel are shown in Table 4 (initial values of electric capacitance calculated for different substances are marked with bold print).

Table 3. Capacitance values for the sensors in the reference channel, with 0.01% variation.

W		$\varepsilon_n = 2.0$			$\varepsilon_n = 2.5$			$\varepsilon_n = 3.0$			$\varepsilon_n = 3.5$	
0%	C_{1-} 29.7 C_{2-} 99	C_1 **30.0** C_2 **100**	C_{1+} 30.3 C_{2+} 101	C_{1-} 37.125 C_{2-} 123.75	C_1 **37.5** C_2 **125**	C_{1+} 37.875 C_{2+} 126.25	C_{1-} 44.55 C_{2-} 148.5	C_1 **45.0** C_2 **150**	C_{1+} 45.45 C_{2+} 151.5	C_{1-} 51.975 C_{2-} 173.25	C_1 **52.5** C_2 **175**	C_{1+} 53.025 C_{2+} 176.75

Table 4. Capacitance values for the sensors in the measuring channel, with 0.01% variation.

ε_n	Variation of C	W = 0%	W = 10%	W = 20%	W = 30%
2.0	C_{3-} C_3 C_{3+}	99 **100** 101	129.393 **130.7** 132.007	166.716 **168.4** 170.084	213.692 **215.85** 218.009
2.5	C_{3-} C_3 C_{3+}	123.75 **125** 126.25	160.974 **162.6** 164.226	206.564 **208.65** 210.737	263.538 **266.2** 268.862
3.0	C_{3-} C_3 C_{3+}	148.5 **150** 151.5	192.308 **194.25** 196.193	245.669 **248.15** 250.632	312.098 **315.25** 318.403
3.5	C_{3-} C_3 C_{3+}	148.5 **150** 151.5	192.308 **194.25** 196.193	245.669 **248.15** 250.632	312.098 **315.25** 318.403
2.0	C_{4-} C_4 C_{4+}	29.7 **30.0** 30.3	38.818 **39.21** 39.602	50.015 **50.52** 51.025	64.107 **64.755** 65.403
2.5	C_{4-} C_4 C_{4+}	37.125 **37.5** 37.875	48.292 **48.78** 49.268	61.969 **62.595** 63.221	79.061 **79.86** 80.659
3.0	C_{4-} C_4 C_{4+}	44.55 **45.0** 45.45	57.692 **58.275** 58.858	73.701 **74.445** 75.189	93.629 **94.575** 95.521

Table 4. Cont.

ε_n	Variation of C	W = 0%	W = 10%	W = 20%	W = 30%
3.5	C_{4-}	51.975	67.003	85.254	107.841
	C_4	**52.5**	**67.68**	**86.115**	**108.93**
	C_{4+}	53.025	68.357	86.976	110.019

The robustness of the transfer function (2) was checked for the conditions when each value of sensors' capacitance was the maximum or minimum within the symmetric variation range ±0.01%. The signs "+" and "−" in later designations represent the maximal or minimal values of electric capacitance. As we had four sensors, 16 different combinations were possible (Table 5). For example, the designation "+−++" means that the electric capacitances C_1, C_3, and C_4 take maximal values, while C_2 is minimal [40].

Table 5. Calculated moisture values with 0.01% variation of capacitances C_1, C_2, C_3, and C_4.

W	$\varepsilon_n = 2.0$ $W_{nominal}$, %				$\varepsilon_n = 2.5$ $W_{nominal}$, %			
	0	10	20	30	0	10	20	30
W_{--++}	0.952	11.038	21.7	31.325	0.952	10.845	21.325	30.884
W_{+-++}	1.252	11.392	22.068	31.601	1.252	11.198	21.694	31.17
W_{-+++}	−0.017	9.89	20.491	30.378	−0.017	9.7	20.115	29.908
W_{++++}	0.268	10.23	20.851	30.666	0.268	10.039	20.475	30.204
W_{----+}	−0.026	9.881	20.481	30.37	−0.026	9.69	20.105	30.204
W_{+---+}	0.268	10.23	20.851	30.666	0.952	10.845	21.325	30.884
W_{-+-+}	−0.973	8.749	19.268	29.364	1.252	11.198	21.694	31.17
W_{++-+}	−0.694	9.083	19.628	29.668	−0.017	9.7	20.115	29.908
W_{--+-}	1.243	11.382	22.057	30.666	0.268	10.039	20.475	30.204
W_{+-+-}	1.546	11.737	22.424	30.956	−0.026	9.69	20.105	30.204
W_{-++-}	0.268	10.23	20.85	30.666	0.268	10.039	20.475	30.504
W_{+++-}	0.556	10.57	21.209	30.948	−0.973	8.561	18.892	29.19
W_{----}	0.268	10.23	20.85	30.666	-0.694	8.895	19.252	29.497
W_{+---}	0.564	10.58	21.22	30.956	1.243	11.188	21.684	31.162
W_{-+--}	−0.686	9.093	19.639	29.677	1.546	11.543	22.052	31.442
W_{++--}	−0.405	9.429	19.999	29.977	0.268	10.039	20.475	30.204
W_{--++}	0.952	10.665	20.945	30.439	0.952	10.473	20.583	29.993
W_{+-++}	1.252	11.017	21.315	30.733	1.252	10.833	20.952	30.296
W_{-+++}	−0.017	9.522	19.734	29.437	−0.017	9.341	19.37	28.968
W_{++++}	0.268	9.86	20.094	29.74	0.268	9.678	19.731	29.278
W_{---+}	−0.026	9.512	19.724	29.428	−0.026	9.331	19.36	28.959
W_{+--+}	0.268	9.86	20.094	29.74	0.268	9.678	19.731	29.278
W_{-+-+}	−0.973	8.386	18.511	28.377	−0.973	8.208	18.147	27.888
W_{++-+}	−0.694	8.719	18.871	28.694	−0.694	8.54	18.507	28.211
W_{--+-}	1.243	11.027	21.304	30.725	1.243	10.823	20.942	30.288
W_{+-+-}	1.546	11.381	21.673	31.015	1.546	11.176	21.311	30.585
W_{-++-}	0.268	9.86	20.094	29.74	0.268	9.678	19.731	29.278
W_{+++-}	0.556	10.199	20.454	30.039	0.556	10.017	20.091	29.583
W_{----}	0.268	9.86	20.094	29.74	0.268	9.678	19.731	29.278
W_{+---}	0.564	10.209	20.464	30.047	0.564	10.026	20.101	29.592
W_{-+--}	−0.686	8.728	18.881	28.703	−0.686	8.55	18.518	28.22
W_{++--}	−0.405	9.063	19.241	29.016	−0.405	8.883	18.878	28.539

As we can see from Table 5, it would be nearly impossible to use separate random results of capacitance measurements together with Formula (2), because the relative full-scale error of moisture content would be rather large. Here, it would be rational to use such robust estimators as the mean (average), median, and truncated mean—in other words, to increase the accuracy of the moisture measurements, we would have to use some kind of estimator, the uncertainty of which would be many times smaller compared with the uncertainty of separate measurements.

Measured values of capacitances for the sensors C_1, C_2, C_3, and C_4 are given in Table 6.

Calculated values of moisture content with mean and median values of the sensors' capacitances for the transfer functions (1) and (2) can be found in Table 7.

To compare the robustness of the results in Table 7, it would be rational to use dispersion analysis. At first, four values of mean square error (dispersions of repeatability) should be calculated (4):

$$S_{rep}^2 = \frac{\sum_{j=1}^{m} \sum_{i=1}^{n} (W_{i,j} - \overline{W_j})^2}{n \cdot m - 1}. \tag{4}$$

Formula (4) requires the mean values of moisture content to be calculated for the results in Table 7. Appropriate calculations are given in Table 8.

After the application of Formula (4), we receive:

$$S_{rep1}^2 = \frac{(0.000-0.000)^2+(9.297-9.151)^2+(19.823-19.393)^2+(31.453-31.657)^2}{2\cdot 4-1} +$$
$$+ \frac{(0.000-0.000)^2+(9.005-9.151)^2+(18.962-19.393)^2+(31.861-31.657)^2}{2\cdot 4-1} = 0.0708,$$

$$S_{rep2}^2 = \frac{(0.268-0.268)^2+(9.218-9.119)^2+(19.687-19.303)^2+(31.266-31.578)^2}{2\cdot 4-1} +$$
$$+ \frac{(0.268-0.268)^2+(9.019-9.119)^2+(18.918-19.303)^2+(31.889-31.578)^2}{2\cdot 4-1} = 0.07279,$$

$$S_{rep3}^2 = \frac{(0.268-0.268)^2+(10.784-10.628)^2+(21.080-20.699)^2+(29.708-29.828)^2}{2\cdot 4-1} +$$
$$+ \frac{(0.268-0.268)^2+(10.472-10.628)^2+(20.318-20.699)^2+(29.948-29.828)^2}{2\cdot 4-1} = 0.0525,$$

$$S_{rep4}^2 = \frac{(0.268-0.268)^2+(10.700-10.594)^2+(20.961-20.620)^2+(29.597-29.781)^2}{2\cdot 4-1} +$$
$$+ \frac{(0.268-0.268)^2+(10.487-10.594)^2+(20.278-20.620)^2+(29.964-29.781)^2}{2\cdot 4-1} = 0.0462,$$

where S_{rep1}^2 is a dispersion of repeatability for the transfer function (1) with mean values of C_1, C_2, C_3, and C_4; S_{rep2}^2 is a dispersion of repeatability for the transfer function (1) with median values; S_{rep3}^2 is a dispersion of repeatability for the transfer function (2) with mean values; and S_{rep4}^2 is a dispersion of repeatability for the transfer function (2) with median values.

Then four dispersions of adequacy (for the data in Table 8) should be calculated as follows (5):

$$S_{ad}^2 = \frac{\sum_{j=1}^{m} (\overline{W_j} - \overline{W}_{jnom})^2}{N - (m+1)}, \tag{5}$$

where N is the total number of moisture content values $W_{calculated}$; and m is the number of nominal points of moisture content W_{nom}.

$$S_{ad1}^2 = \frac{(0-0)^2+(9.151-10)^2+(19.393-20)^2+(31.657-30)^2}{8-(4+1)} = 1.302,$$

$$S_{ad2}^2 = \frac{(0-0)^2+(9.119-10)^2+(19.303-20)^2+(31.578-30)^2}{8-(4+1)} = 1.275,$$

$$S_{ad3}^2 = \frac{(0.268-0)^2+(10.628-10)^2+(20.699-20)^2+(29.828-30)^2}{8-(4+1)} = 0.328,$$

$$S_{ad4}^2 = \frac{(0.268-0)^2+(10.594-10)^2+(20.620-20)^2+(29.781-30)^2}{8-(4+1)} = 0.286,$$

where S_{ad1}^2 is a dispersion of adequacy for the transfer function (1) with mean values; S_{ad2}^2 is a dispersion of adequacy for the transfer function (1) with median values; S_{ad3}^2 is a dispersion of adequacy for the transfer function (2) with mean values; and S_{ad4}^2 is a dispersion of adequacy for the transfer function (2) with median values.

To compare the dispersions of adequacy with the dispersions of repeatability, we used the following F-test: $F = S_{ad}^2 / S_{rep}^2$. All calculations for the results given in Tables 7 and 8 can be found in Table 9.

Table 6. Measured values of capacitance for the sensors C_1, C_2, C_3, and C_4 with diesel fuel and mazut.

№ of Experiment	Moisture Content W, %							
	0				10		20	30
	Diesel Fuel, $\varepsilon_n = 2.01$							
	C_1, pF	C_2, pF	C_3, pF	C_4, pF	C_3, pF	C_4, pF	C_3, pF	C_4, pF
1	30.22	94.69	125.34	40.62	159.45	51.24	202.91	68.36
2	30.61	94.81	125.21	40.30	160.11	50.96	203.26	68.24
3	30.29	95.16	125.64	40.36	159.66	50.84	203.11	67.86
4	30.48	94.73	125.82	40.29	159.74	51.11	202.84	67.91
5	30.41	94.79	125.70	40.32	159.91	51.03	202.89	67.99
6	30.36	94.91	125.49	40.59	159.86	50.91	203.04	68.02
7	30.28	94.98	125.66	40.44	159.72	50.89	203.13	68.17
8	30.31	95.11	125.21	40.26	159.84	51.36	202.94	68.06
9	30.42	94.73	125.32	40.32	159.79	51.01	202.63	68.24
10	30.41	94.85	125.61	40.41	160.03	51.26	202.79	67.93
Mean values	30.58	94.81	125.50	40.39	159.81	51.06	202.95	68.08
Median values	30.39	94.83	125.55	40.34	159.82	51.02	202.93	68.04
№ of Experiment	Mazut, $\varepsilon_n = 2.67$							
	C_1, pF	C_2, pF	C_3, pF	C_4, pF	C_3, pF	C_4, pF	C_3, pF	C_4, pF
1	41.16	127.56	168.46	54.08	213.86	70.36	270.64	87.52
2	41.29	127.64	168.04	54.21	214.21	70.24	270.99	88.26
3	41.36	128.12	168.54	54.66	214.10	70.08	271.04	87.96
4	41.28	127.74	168.22	54.71	214.36	70.21	270.45	87.71
5	41.40	127.81	168.11	54.32	214.02	69.96	270.93	87.82
6	41.31	127.94	168.01	54.39	213.69	69.91	270.44	87.46
7	41.19	127.90	168.14	54.56	213.76	70.22	270.51	87.56
8	41.22	127.69	168.34	54.38	213.84	70.16	270.72	87.39
9	41.38	127.78	168.21	54.49	213.92	70.28	270.59	87.94
10	41.31	128.09	168.03	54.41	213.90	69.08	270.41	87.62
Mean values	41.29	127.83	168.21	54.42	213.97	70.05	270.67	87.72
Median values	41.30	127.80	168.18	54.40	213.91	70.19	270.62	87.67

Table 7. Calculated results of the transfer functions (1) and (2), with mean and median values.

W_{nom}, %	$W_{calculated}$, %			
	Static Function (1)		Static Function (2)	
	with Mean Values of C_1, C_2, C_3 and C_4		with Mean Values of C_1, C_2, C_3 and C_4	
	Diesel Fuel	Mazut	Diesel Fuel	Mazut
0	0.000	0.000	0.268	0.268
10	9.297	9.005	10.784	10.472
20	19.823	18.962	21.080	20.318
30	31.453	31.861	29.708	29.948
	with Median Values of C_1, C_2, C_3 and C_4		with Median Values of C_1, C_2, C_3 and C_4	

Table 7. Cont.

W_{nom}, %	$W_{calculated}$, %			
	Static Function (1)		Static Function (2)	
	with Mean Values of C_1, C_2, C_3 and C_4		with Mean Values of C_1, C_2, C_3 and C_4	
	Diesel Fuel	Mazut	Diesel Fuel	Mazut
0	0.000	0.000	0.268	0.268
10	9.218	9.019	10.700	10.487
20	19.687	18.918	20.961	20.278
30	31.266	31.889	29.597	29.964

Table 8. Arithmetic mean values of moisture content.

W_{nom}, %	Mean Values $\overline{W_j}$, %			
	Function (1), Mean Values	Function (1), Median Values	Function (2), Mean Values	Function (2), Median Values
0	0.000	0.000	0.268	0.268
10	9.151	9.119	10.628	10.594
20	19.393	19.303	20.699	20.620
30	31.657	31.578	29.828	29.781

Table 9. Results of robustness verification for the transfer functions (1) and (2).

Data Source	S_{rep}^2	S_{ad}^2	F
Function (1), mean values	0.0708	1.302	18.39
Function (1), median values	0.0728	1.275	17.51
Function (2), mean values	0.0525	0.328	6.25
Function (2), median values	0.0462	0.286	6.19

4. Results/Discussion

As mentioned in the "Materials and Methods" section, the first attempt to provide effective "type uncertainty" compensation for capacitive moisture meters was described in [8], and transfer function (1) is the result.

The calculated values of the sensors' capacitances and moisture content for the transfer function (1) were received with a help of the universal Wiener equation as one of the most popular dielectric mixture models (6):

$$\varepsilon_{mix} = \varepsilon_n \left(1 + \frac{3W}{(\varepsilon_W + 2\varepsilon_n)/(\varepsilon_W - \varepsilon_n) - W} \right), \quad (6)$$

where ε_{mix} is the dielectric permittivity of a binary material–water mixture; ε_n is the dielectric permittivity of dry material; $\varepsilon_W = 80$, and is the dielectric permittivity of water; and W is the volumetric moisture content, taken in absolute values.

If we apply the universal Wiener equation to the values of dielectric permittivity given in Table 1, and use it as the water-cut meter's transfer function, we obtain a family of curves (Figure 9).

Figure 9 illustrates the situation when the oil matrix composition changes as well as the water content (oil type uncertainty). If no type-uncertainty compensation happens, we have a family of transfer functions for the water-cut meter instead of one, taken as a nomi-nal.

The transfer function (1) helps to solve this problem, and provides the possibility of measurements in emulsified fuels, where the dielectric permittivity of a dry oil ε_n varies in a wide range from 2 to 3.5, without any preliminary calibration.

However, Figure 1 illustrates that a transfer function received with the help of Equation (1) is still nonlinear, the influence of fuel type uncertainty remains significant, and the calculated values of moisture content do not correspond to the nominal linear model.

To determine the linear dependence between moisture content W and the $(C_3 - C_4)/(C_2 - C_1)$ relationship, we used the method of least squares:

$$a + b \cdot W = \frac{C_3 - C_4}{C_2 - C_1}, \qquad (7)$$

where a and b are unknown coefficients of a first-order polynomial (7).

To build a system of conditional equations, we had to choose average values of the $(C_3 - C_4)/(C_2 - C_1)$ relationship (Table 10).

A system of linear conditional equations, based on the first-order polynomial (7), is given below.

$$\begin{cases} a + b \cdot 0 = 1, \\ a + b \cdot 10 = 1.298, \\ a + b \cdot 20 = 1.662, \\ a + b \cdot 30 = 2.116. \end{cases}$$

It was solved in a traditional way:

$$Q = \begin{vmatrix} [XX] & [XY] \\ [YX] & [YY] \end{vmatrix} = \begin{vmatrix} 4 & 60 \\ 60 & 1400 \end{vmatrix} = 2000, Q_a = \begin{vmatrix} [XL] & [XY] \\ [YL] & [YY] \end{vmatrix} = \begin{vmatrix} 6.076 & 60 \\ 109.7 & 1400 \end{vmatrix} = 1924.4,$$

$$Q_b = \begin{vmatrix} [XX] & [XL] \\ [YX] & [YL] \end{vmatrix} = \begin{vmatrix} 4 & 6.076 \\ 60 & 109.7 \end{vmatrix} = 74.24,$$

$$a = \frac{Q_a}{Q} = \frac{1924.4}{2000} = 0.9622, \quad b = \frac{Q_b}{Q} = \frac{74.24}{2000} = 0.0371.$$

Table 10. Average values for the $(C_3 - C_4)/(C_2 - C_1)$ relationship.

W_{nom}, %	$(C_3 - C_4)/(C_2 - C_1)$				
	$\varepsilon_n = 2.0$	$\varepsilon_n = 2.5$	$\varepsilon_n = 3.0$	$\varepsilon_n = 3.5$	Average
0	1	1	1	1	1
10	1.307	1.301	1.295	1.289	1.298
20	1.684	1.669	1.654	1.640	1.662
30	2.158	2.130	2.102	2.075	2.116

Calculated values of moisture W_{calc}, received from a first-order polynomial (7):

$$W_{calc} = \frac{C_3 - C_4}{(C_2 - C_1) \cdot 0.0371} - 25.9353,$$

are given in Table 11.

Table 11. Values of moisture content W' obtained from a first-order polynomial (7).

W_{nom}, %	W_{calc}, %			
	$\varepsilon_n = 2.0$	$\varepsilon_n = 2.5$	$\varepsilon_n = 3.0$	$\varepsilon_n = 3.5$
0	1.019	1.019	1.019	1.019
10	9.294	9.132	8.970	8.809
20	19.456	19.051	18.647	18.270
30	32.232	31.477	30.722	29.995

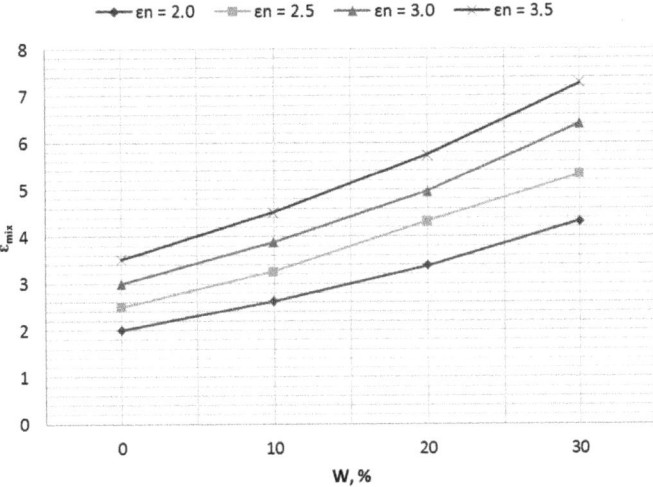

Figure 9. Influence of the changes in the composition of a dry substance ε_n on the value of the dielectric permittivity of a binary mixture ε_{mix}.

After obtaining the results of Table 11, it was possible to conclude that the least squares method is not effective for our case, as it was expected to obtain the linear transfer function of the water-cut meter (Figure 10). The next step of the linearization was to calculate the discrepancies ΔW between nominal points of moisture content and values from Table 11, and to approximate them by applying the instruments of general linear regression. Discrepancies between nominal points of moisture content and values from Table 11, taken for $\varepsilon_n = 3.0$, are given in Table 12.

Table 12. Values of moisture content W' obtained from a first-order polynomial (7).

	$\varepsilon_n = 3.0$			
$W_{nominal}$, %	0	10	20	30
W_{calc}, %	1.019	8.970	18.647	30.722
$\Delta W = W_{calc} - W_{nominal}$, %	1.019	−1.030	−1.353	0.722

To approximate the values of discrepancies ΔW from Table 11, a sum of four functions (1, W, W^2, and W^3) was taken with appropriate coefficients, defined with the help of Mathcad software (linfit function (x, y, Y)):

$$\Delta W = 1.019 - 0.269 \cdot W + 0.00527 \cdot W^2 + 0.000112 \cdot W^3, \tag{8}$$

where W is a volumetric moisture content taken from the first-order polynomial (7). Formula (8) provides an ideal approximation of the discrepancies ΔW, as can be seen in Table 13.

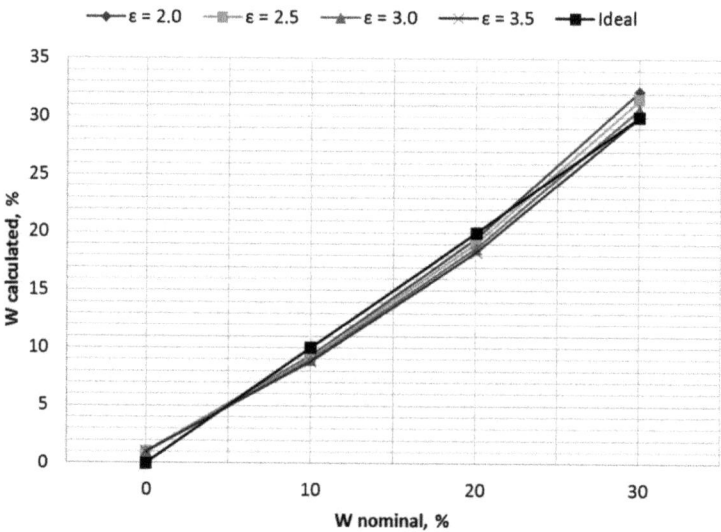

Figure 10. Graphs of a first-order polynomial (7), placed together with an ideal transfer function of the water-cut meter.

Table 13. Approximated values of discrepancies ΔW_{calc}.

	$\varepsilon_n = 3.0$			
ΔW	1.019	−1.030	−1.353	0.722
ΔW_{calc}	1.019	−1.030	−1.353	0.722

Taking into account Equations (7) and (8), we received a modified transfer function of a water-cut meter (function 2), as displayed in Figure 5.

During the comparison of the water-cut meter's transfer functions (1) and (2), it was necessary to estimate how both of them compensated for the type uncertainty, and how close both of them were to the nominal linear transfer function. Dispersions of repeatability, calculated for transfer function (1) and transfer function (2) with mean and median values, describe a possible variation in the measured moisture values for two chosen substances (diesel fuel and mazut). We can see that the values of S^2_{rep1} and S^2_{rep2} are approximately 1.35 times greater in comparison with S^2_{rep3} and S^2_{rep4}. This means that transfer function (2) provides a lower level of variability of the measured values, depending on the type of oil fuel under research. In other words, transfer function (2) is not as sensitive to fuel type as transfer function (1).

Dispersions of adequacy reflect the correspondence of experimental transfer functions (1) and (2) (taken for diesel fuel and mazut as dielectric substances with a significant difference in relative permittivity) with the nominal linear transfer function of a water-cut meter (the ideal transfer function given in Figures 4 and 5). Again, it is possible to see that dispersions of adequacy calculated for the transfer function (2) give us values that are approximately 4.5 times smaller in comparison with the dispersions of transfer function (1). This means that Equation (2) provides 4.5 times better adequacy to the ideal than Equation (1).

As a general result, we can conclude that it would be rational to choose transfer function (2) with median values of moisture content as a nominal transfer function of the water-cut meter, because it has the smallest F-test value ($F = 6.19$).

Moisture measurement uncertainty can be estimated by using the method of mathematical programing [41]. If the function $f(x)$ is continuous, it is possible to find maximal and minimal values of $f(x)$ within the limit value of the function's argument error. Absolute uncertainty of moisture measurement can be calculated as a half difference $\Delta f = (f_{max}(x) - f_{min}(x))/2$. Maximal and minimal capacitance values were taken from the results of 10 random measurements in mazut with 30% moisture content: C_{1min} = 41.16 pF, C_{1max} = 41.40 pF, C_{2min} = 127.56 pF, C_{2max} = 128.09 pF, C_{3min} = 270.41 pF, C_{3max} = 271.04 pF, C_{4min} = 87.39 pF, and C_{4max} = 88.26 pF. After substituting these capacitances into Formula (2), we received the following maximal and minimal values of moisture content: W_{max} = 30.235%, and W_{min} = 29.633%. The value of absolute moisture measurement's extended uncertainty was equal to $U(W)$ = (30.235 − 29.633)/2 = 0.3%, which is very good for a capacitive water-cut meter, because modern accuracy limits are mostly no lower than 0.5% (in [20,24], we have 1%; in [26], −0.67%; in [21], −3%...)

5. Conclusions

At first, the workability of a transfer function suggested in [17] with different types of oil fuels was checked. Four values of dielectric permittivity for dehydrated substances (ε_n = 2.0; ε_n = 2.5; ε_n = 3.0; ε_n = 3.5) and four values of moisture content (W = 0%; W = 10%; W = 20%; W = 30%) were taken for that purpose. The obtained graphs of transfer functions for the mentioned substances happened to be nonlinear, and the calculated values of moisture content were significantly different from the nominal at all points, except for W = 0%. To achieve near-linear transfer function of the capacitive water-cut meter, the method of least squares was used. The next step of linearization was in calculating discrepancies between the nominal points of moisture content and values received after the application of the LS method, with their approximation using general linear regression instruments.

Analyzing the graphs of the modified transfer functions for different moist substances, it was possible to say that they happened to be far more effective than the initial transfer function suggested in [17]. The root mean estimator was calculated for the initial transfer function, and the static function—received after general linear regression—as an integral difference between the nominal and calculated values of moisture content. The corresponding root mean estimator values were 1.3062% and 0.4158%, proving the effectiveness of transfer function (2), modified with the help of general linear regression instruments.

To check the workability of the modified transfer function in real conditions, two different fuels were taken: diesel fuel and mazut. The values of moisture content for two transfer functions, calculated for the mean and median values of C_1, C_2, C_3, and C_4, were compared with the help of dispersion analysis. After the dispersions of repeatability and adequacy had been calculated and compared, it was possible to conclude that the modified transfer function of a water-cut meter, calculated for median values of C_1, C_2, C_3, and C_4, provided minimal dispersions of repeatability and adequacy, and a minimal F-test value. Values of both dispersions for the modified transfer function were significantly smaller, as were the F-test values, proving its more robust properties (i.e., capacity for type-uncertainty compensation) and better adequacy to the nominal linear transfer function of the water-cut meter.

Author Contributions: Conceptualization, O.Z. and N.K.; methodology, O.Z.; validation, O.Z. and V.Z.; formal analysis, O.Z. and N.K.; investigation, O.Z. and V.Z.; resources, V.Z.; data curation, O.Z. and N.K.; writing—original draft preparation, O.Z.; writing—review and editing, N.K.; visualization, V.Z.; supervision, O.Z.; project administration, O.Z.; funding acquisition, O.Z. and V.Z. All authors have read and agreed to the published version of the manuscript.

Funding: This research received no external funding.

Conflicts of Interest: The authors declare no conflict of interest.

References

1. Girin, O. Dynamics of the emulsified fuel drop microexplosion. *At. Sprays* **2017**, *27*, 407–422. [CrossRef]
2. Abdollahi, M.; Ghobadian, B.; Najafi, G.; Hoseini, S.S.; Mofijur, M.; Mazlan, M. Impact of water–biodiesel–diesel nano-emulsion fuel on performance parameters and diesel engine emission. *Fuel* **2020**, *280*, 118576. [CrossRef]
3. Kadota, T.; Yamasaki, H. Recent advances in the combustion of water fuel emulsion. *Prog. Energy Combust. Sci.* **2002**, *28*, 385–404. [CrossRef]
4. Allaa, M.; Soulayman, S.; Abdelkarim, T.; Walid, Z. Water/Heavy fuel oil emulsion production, characterization and combustion. *Int. J. Renew. Energy Dev.* **2021**, *10*, 597–605. [CrossRef]
5. Gopidesi, R.K.; Rajaram, P.S. A review on emulsified fuels and their application in diesel engine. *Int. J. Ambient. Energy* **2019**, *43*, 732–740. [CrossRef]
6. Badran, O.; Emeish, S.; Abu-Zaid, M.; Abu-Rahma, T.; Al-Hasan, M.; Al-Ragheb, M. Impact of Emulsified Water/Diesel Mixture on Engine Performance and Environment. *Int. J. Therm. Environ. Eng.* **2011**, *3*, 1–7. [CrossRef]
7. Woo, S.; Kim, W.; Lee, J.; Lee, K. Fuel properties of water emulsion fuel prepared using porous membrane method for low pollutant engine at various temperatures. *Energy Rep.* **2021**, *7*, 6638–6650. [CrossRef]
8. Vellaiyan, S.; Amirthagadeswaran, K.S. The role of water-in-diesel emulsion and its additives on diesel engine performance and emission levels: A retrospective review. *Alex. Eng. J.* **2016**, *55*, 2463–2472. [CrossRef]
9. Sugeng, D.A.; Ithnin, A.M.; Yahya, W.J.; Kadir, H.A. Emulsifier-free water-in-biodiesel emulsion fuel via steam emulsification: Its physical properties, combustion performance, and exhaust emission. *Energies* **2020**, *13*, 5406. [CrossRef]
10. Salih, S.Y. Water-diesel emulsion: A review. *Int. J. Adv. Eng. Technol.* **2017**, *10*, 429–436.
11. Zabolotnyi, O.; Zabolotnyi, V.; Koshevoy, N. Oil products moisture measurement using adaptive capacitive instrument measuring transducers. In *Lecture Notes in Networks and Systems, Integrated Computer Technologies in Mechanical Engineering-2020*; Nechyporuk, M., Pavlikov, V., Kritskiy, D., Eds.; Springer: Cham, Switzerland, 2020; Volume 188, pp. 81–89. [CrossRef]
12. Khanjani, A.; Sobati, M.A. Performance and emission of a diesel engine using different water/waste fish oil (WFO) biodiesel/diesel emulsion fuels: Optimization of fuel formulation via response surface methodology (RSM). *Fuel* **2021**, *288*, 119662. [CrossRef]
13. Suryadi, J.; Winardi, S.; Nurtono, T. The effect of water contents to diesel fuel-water emulsion fuel stability. *IPTEK J. Technol. Sci.* **2019**, *30*, 28–31. [CrossRef]
14. Kane, J.M.; Prat-Guitart, N. Fuel moisture. In *Encyclopedia of Wildfires and Wildland-Urban Interface (WUI) Fires*; Manzello, S., Ed.; Springer: Cham, Switzerland, 2018. [CrossRef]
15. Sharma, P.; Yeung, H. Recent advances in water cut sensing technology. In *Advances in Measurements and Instrumentation: Reviews*; Yuris, S.Y., Ed.; IFSA Publishing: Barselona, Spain, 2018; Volume 1, pp. 147–170.
16. Teixeira, C.E.; da Silva, L.E.B.; Veloso, G.F.; Lambert-Torres, G.; Campos, M.M.; Noronha, I.; Bonaldi, E.L.; de Oliveira, L.E.L. An ultrasound-based water-cut meter for heavy fuel oil. *Measurement* **2019**, *148*, 106907. [CrossRef]
17. Zabolotnyi, O.; Koshevoi, M. An effective method of bulk materials moisture measurement using capacitive sensors. *J. Stored Prod. Res.* **2020**, *89*, 101733. [CrossRef]
18. Königer, F.; Stacheder, M.; Kaden, H.; Schuhmann, R. Electrophysical principles of moisture measurement. In Proceedings of the Innovative Feuchtemessung in Forschung und Praxis: Materialeigenschaften und Prozesse, Berichtsband zur Tagung, Karlsruhe, Germany, 12–14 October 2011.
19. Yang, Q.; Pei, X.; Liu, F.; Wang, L.; Yan, J.; Wang, J. An electromagnetic wave based water cut meter and the influence factors of measuring accuracy. In Proceedings of the SPE/IATMI Asia Pacific Oil & Gas Conference, Nusa Dua, Bali, Indonesia, 22–25 October 2015. [CrossRef]
20. Deng, X.; Yang, L.; Fu, Z.; Du, C.; Lyu, H.; Cui, L.; Zhang, L.; Zhang, J.; Jia, B. A calibration-free capacitive moisture detection method for multiple soil environments. *Measurement* **2021**, *173*, 108599. [CrossRef]
21. Osadchuk, O.V.; Semenov, A.O.; Zviahin, O.S.; Semenova, O.O.; Rudyk, A.V. Increasing the sensitivity of measurement of a moisture content in crude oil. *Nauk. Visnyk Natsionalnoho Hirnychoho Universytetu* **2021**, *5*, 49–53. [CrossRef]
22. Aslam, M.Z.; Tang, T.B. A high resolution capacitive sensing system for the measurement of water content in crude oil. *Sensors* **2014**, *14*, 11351–11361. [CrossRef]
23. Campos, M.M.; Borges-da-Silva, L.E.; Arantes, D.d.A.; Teixeira, C.E.; Bonaldi, E.L.; Lambert-Torres, G.; Ribeiro Junior, R.F.; Krupa, G.P.; Sant'Ana, W.C.; Oliveira, L.E.L.; et al. An ultrasonic-capacitive system for online characterization of fuel oils in thermal power plants. *Sensors* **2021**, *21*, 7979. [CrossRef]
24. Corach, J.; Galván, E.F.; Sorichetti, P.A.; Romano, S.D. Broadband permittivity sensor for biodiesel and blends. *Fuel* **2019**, *254*, 115679. [CrossRef]
25. Vello, T.P.; de Oliveira, R.F.; Silva, G.O.; de Camargo, D.H.S.; Bufon, C.C.B. A simple capacitive method to evaluate ethanol fuel samples. *Sci. Rep.* **2017**, *7*, 43432. [CrossRef]
26. Delfino, J.R.; Pereira, T.C.; Costa Viegas, H.D.; Marques, E.P.; Pupim Ferreira, A.A.; Zhang, L.; Zhang, J.; Brandes Marques, A.L. A simple and fast method to determine water content in biodiesel by electrochemical impedance spectroscopy. *Talanta* **2018**, *179*, 753–759. [CrossRef] [PubMed]

27. Hegg, M.C.; Mamishev, A.V. Influence of variable plate separation on fringing electric fields in parallel-plate capacitors. In Proceedings of the IEEE International Symposium on Electrical Insulation, Indianapolis, IN, USA, 19–22 September 2004. [CrossRef]
28. Zabolotny, A.V.; Koshevoi, M.D. Improving efficiency of the quality control of substances with dielectric properties. *Telecommun. Radio Eng.* **2002**, *57*, 177–190. [CrossRef]
29. Hanai, T.; Koizumi, N.; Goto, R.; Koizumi, N.; Goto, R. Dielectric Constants of Emulsions. *Bull. Inst. Chem. Res.* **1962**, *40*, 240–271.
30. Wobschall, D. A theory of the complex dielectric permittivity of soil containing water: The semidisperse model. *IEEE Trans. Geosci. Electron.* **1977**, *15*, 49–58. [CrossRef]
31. Nandi, N.; Bhattacharyya, K.; Bagchi, B. Dielectric relaxation and solvation dynamics of water in complex chemical and biological systems. *Chem. Rev.* **2000**, *100*, 2013–2046. [CrossRef]
32. Goncharenko, A.V.; Lozovski, V.Z.; Venger, E.F. Lichtenecker's equation: Applicability and limitations. *Opt. Commun.* **2000**, *174*, 19–32. [CrossRef]
33. Zabolotnyi, O. Moisture content control in heavy fuel during the process of emulsification with a help of capacitive sensors. In Proceedings of the 25th International Scientific Conference 'Transport Means 2021', Part 1, Kaunas, Lithuania, 6–8 October 2021.
34. *ISO 835:2007*; Laboratory Glassware–Graduated Pipettes. ISO: Geneva, Switzerland, 2007; 12 p.
35. Li, Y.; Feng, Y.; Yu, G.; Li, J.; Zhou, Y.; Liu, Y. Preparation and characterization of oil-in-water emulsion based on eco-friendly emulsifiers. *Colloids Surf. A Physicochem. Eng. Asp.* **2020**, *602*, 125024. [CrossRef]
36. Chen, G.; Tao, D. An experimental study of stability of oil-water emulsion. *Fuel Process. Technol.* **2005**, *86*, 499–508. [CrossRef]
37. Apley, D.W.; Kim, J. A cautious approach to robust design with model parameter uncertainty. *IIE Trans.* **2011**, *43*, 471–482. [CrossRef]
38. Marandi, A.; den Hertog, D. When are static and adjustable robust optimization problems with constraint-wise uncertainty equivalent? *Math. Program.* **2018**, *170*, 555–568. [CrossRef]
39. Meniailov, I.; Ugryumov, M.; Chumachenko, D.; Bazilevych, K.; Chernysh, S.; Trofymova, I. Non-linear estimation methods in multi-objective problems of robust optimal design and diagnostics of systems under uncertainties. In *Lecture Notes in Networks and Systems, Integrated Computer Technologies in Mechanical Engineering-2020*; Nechyporuk, M., Pavlikov, V., Kritskiy, D., Eds.; Springer: Cham, Switzerland, 2020; Volume 1113, pp. 198–207. [CrossRef]
40. Zabolotnyi, O.V.; Zabolotnyi, V.A.; Koshevoi, M.D. Conditionality examination of the new testing algorithms for coal-water slurries moisture measurement. *Sci. Bull. Nat. Min. Univ.* **2018**, *1*, 51–59. [CrossRef]
41. Chen, H.-Y.; Lee, A.C.; Lee, C.F. Alternative Methods to Deal with Measurement Error. In *Handbook of Financial Econometrics, Mathematics, Statistics, and Machine Learning*; World Scientific Publishing Co. Pte Ltd.: Singapore, 2020; pp. 1439–1484. [CrossRef]

Article

The Methods of Three-Dimensional Modeling of the Hydrogenerator Thrust Bearing

Oleksii Tretiak [1,*], Dmitriy Kritskiy [2,*], Igor Kobzar [3,*], Mariia Arefieva [4,*] and Viacheslav Nazarenko [1,*]

1. Faculty of Aircraft Engineering, Department of Aerohydrodynamics, National Aerospace University Kharkiv Aviation Institute, 61070 Kharkov, Ukraine
2. Department of Information Technology Design, National Aerospace University Kharkiv Aviation Institute, 61070 Kharkov, Ukraine
3. Special Design Office for Turbogenerators and Hydrogenerators, Join Stock Company "Ukrainian Energy Machines", 61037 Kharkov, Ukraine
4. Kharkiv Lyceum "IT Step School Kharkiv", 61010 Kharkov, Ukraine
* Correspondence: alex3tretjak@ukr.net (O.T.); d.krickiy@khai.edu (D.K.); ivkobzar@ukr.net (I.K.); marii.arefieva@gmail.com (M.A.); my_registrator@ukr.net (V.N.)

Abstract: In the presented scientific work, the basic design versions of the thrust bearings of Hydrogenerators are considered. The main causes of emergencies in the thrust bearing unit of a high-power Hydrogenerator are considered. The main requirements for the operation of thrust bearings are submitted. Cause-and-effect relationships of emerging and development of defects are established. Existing methods for calculating the stressed state of a thrust bearing in the classical formulation for a stationary mode of operation are considered. The main features of the operation of the thrust bearing unit are investigated in relation to the features of the sliding bearings. The calculation of the elastic chambers of the hydraulic thrust bearing in a three-dimensional formulation is carried out, taking into account the physical properties of the oil, the material of the chambers and distribution of the acting loads. It is shown that the applied designs of Join Stock Company "Ukrainian Energy Machines" can be used in high-power Hydrogenerators.

Keywords: the thrust bearing; fatigue calculation; three-dimensional modeling; fatigue curve

1. Introduction

Hydrogenerators like other electrical machines consist of active and constructional parts. Active parts directly involved in to the process of energy conversion including the rotor and stator magnetic cores circuits with windings on them. All other elements are called constructive ones. The construction of the vertical Hydrogenerator is presented in the scientific work of A. I. Abramov [1].

A characteristic feature of the construction of vertical Hydrogenerator is the presence of the support bearing, called the thrust bearing. It reliably operates under the influence of large forces created by the masses of the rotor of Hydrogenerator and the turbine rotor, as well as the vertical component of the water pressure acting on the turbine runner. These forces are transferred through the thrust bearing bush fixed on the shaft and the thrust bearing on the load bearing or supporting spider and then through the stator body to the foundation.

As a rule, the main units of large electrical machines operate with loads, the mechanical stresses of which are significantly removed from the maximum permissible. Therefore, the probability of fatigue failure appears very rarely. However, when they appear, it leads to significant structural damage or to the replacement of existing support structures (the refurbishment time takes an extended period). A significant contribution in to the solution of the problem of determining the fatigue stresses for the main generating equipment was made by Chernousenko O. Yu. [2].

The thrust bearings operate in the oil friction mode. They are characterized by exceptional durability and low friction coefficients.

The calculation of thrust bearings on rigid supports is submitted in the paper [3,4]. A significant advantage of the work is the consideration of the oil wedge. However, given thrust bearings are not used in the design of high-power hydrogenerating units, as they have limitations on the ultimate load on the support planes.

In the paper [5], the types of failures of hydrogenerating units are submitted in detail, however, its usage is limited by the lack of a three-dimensional statement of the problem under consideration.

In the paper [6], the causes of damage due to misalignment of the shafts of hydrogenerating units are indicated, the presented work has limitations in application, due to the fact that for long shafts of hydrogenerating units, the runout is monitored by the control system, and work with unacceptable vibrations is prohibited. In general, this algorithm, specified in the paper [7], can be adapted to large machines, to study the operation after 30 years of operation (over-design mode of operation).

In the paper [8], the vibrations of hydrogenerating units were studied; it should be noted the high reliability of the obtained information. An undoubted advantage is the study of vibrations of a real design structure.

When designing new hydrogenerating units, it is necessary to take into consideration the experience of the paper [9] of Michell Bearings for 65 years of its existence, which combines practical innovations with modern design technologies.

At the same time, when calculating the oil wedge in the bearing surface, it is necessary to consider and pay a special attention to the paper [10], as it allows you to answer the issues of the correct choice of gaskets a prototype set of PTFE pads.

In the paper [11] schematic and power solutions for the thrust bearing of an operating hydrogenerating unit are submitted. In this case, the obtained dependencies were correlated by the author with the experiment. And the paper [12] regulates in detail the assembly process and restrictions on the operation of hydro-electric units. In the paper [13], a research facility was developed and manufactured to study the design and opereational characteristics of industrial hydrodynamic thrust bearings, which allows verifying the results of previously presented scientific studies.

In the paper [14], a team of authors performed a three-dimensional analysis of the stress-strain state of a thrust bearing on rigid supports. At the same time, the presence of a thermoelastic component shall be considered as a significant advantage of the work. However, due to the fact that the thrust bearings on hydraulic supports include vessels under high pressure, this algorithm is not applicable for hydrogenerating units with a power over than 300 MW. The paper [15] deserves a special attention in terms of modeling the flow of the oil layer, while paper [16] can be supplemented with the 3-rd (the third) boundary conditions obtained from the results of the paper [15].

The invention of inclined sliding bearings A. G. M. Michell [16] in Australia and independently, A. Kingsbury [17] in the USA let use existing designs during decades.

However, during the last time there was a sharp jump in technical and design capabilities, which allowed creating qualitatively new products.

In Figure 1 existing designs Waukesha Bearings and Itaipu Power Plant are submitted, where 1 is the thrust bush (skirt), 2 is the shaft of Hydroaggregate, 3 is cooler, 4 is the thrust segment.

One of the options for ensuring the stability of the thrust bearing operation is the introduction of an additional layer into the friction zone (see Figure 2).

However, given design does not solve completely all the technical problems as bronze is electrically conducting material.

The elastic chambers with corrugation are used in the original design of the thrust bearing unit of JSC "Ukrainian Energy Machines". Such design is described in scientific work [18]. Given design solution let provide reliable operation of the thrust bearing unit with loads of 300 t. Due to the fact that Hydroaggregates should operate not less than

40 years and operation conditions demand high maneuverability of the electrical machines then it is necessary to carry out a fatigue analysis of the thrust bearing unit operation for the modes where the service life should overcome the claimed one.

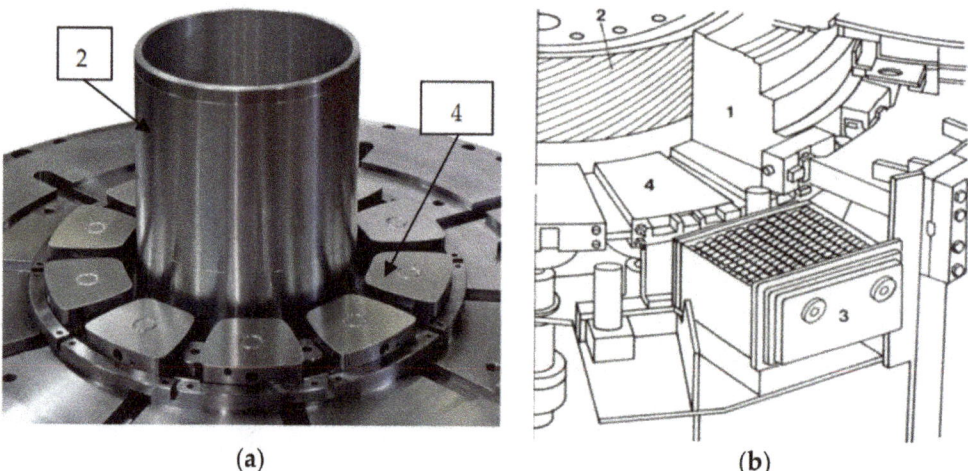

Figure 1. Tilting pad thrust bearing: (a) courtesy of Waukesha Bearings; (b) courtesy of Itaipu Power Plant.

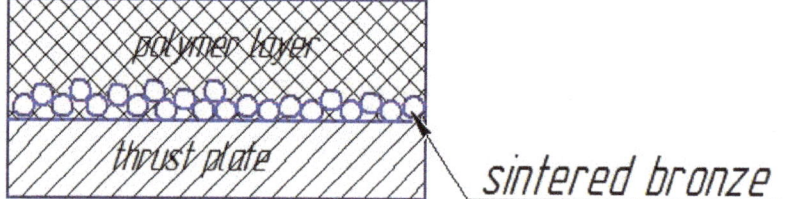

Figure 2. Location of friction with the bronze-plated.

1.1. The Main Causes of Damage of the Thrust Bearing Chambers

The main cause of damage of the thrust bearing chambers of the vertical Hydrogenerator is depressurization of chambers at that the thrust bearing from hydraulic with automatic load balancing between the segments turns into the thrust bearing on rigid supports. However, as a rule, there is a significant misalignment of the load on the segments.

In Figure 3 the Diagram of development of the above-mentioned defect of the thrust bearing on the hydraulic support is shown. The causes of the defect (they are shown at the 1-st and 2-nd time levels) are increased pulsation in the thrust bearing or manufacturing defects. In the time levels 3–7 the further development of the emergency situation is shown.

When cracks appear in the elastic chamber its replacement is necessary followed by filling of whole system with oil. Features of the study of emergency situations are submitted in scientific work [19].

1.2. Main Requirements to the Thrust Bearing Unit of Hydroaggregate Rated 330 MW

At operation of Hydrogenerator-Motor rated 330 MW in the generator mode during the start the axial load on to the thrust bearing from the masses of rotating parts of the pump-turbine and water reaction can achieve of 3000 kN and total axial load from water pressure and the rotating parts of the Hydroaggregate at rated mode shall not exceed 23,000 kN.

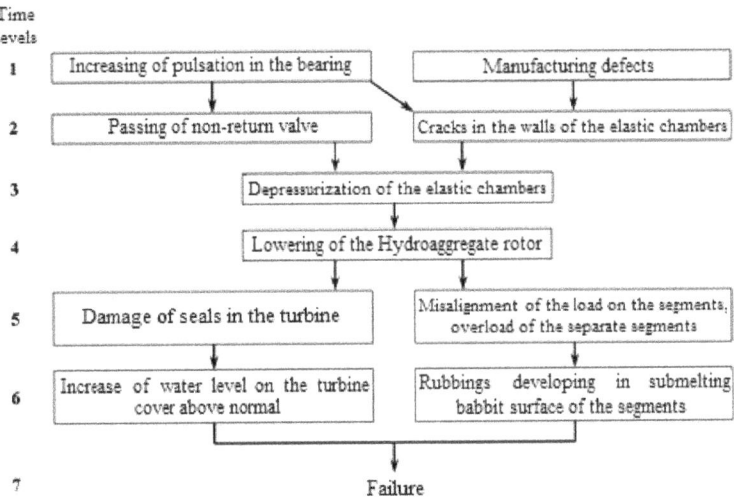

Figure 3. The Diagram of Defects Development of the Thrust Bearing Chambers.

At that the requirements to safety and health protection are strictly regulated by the relevant documents. Hydrogenerator for safety of the design shall comply with GOST 12.1.004-91, GOST 12.1.010-76, GOST 12.2.007.0-75 and GOST 12.2.007.1-75. According to the method of protection of a person from electric current shock, Hydrogenerator has class 01 in accordance with GOST 12.2.007.0-75, and its fire safety corresponds to GOST 12.1.004-91. The degree of protection with shell of the Hydrogenerator-Motor shall be IP00 according to Publication IEC 34-5.

In addition, the vertical vibration of the thrust bearing base (double amplitude) shall not exceed of 0.15 mm, and the temperature of the surface of the Hydrogenerator-Motor during inspections and repairs is limited to 45 °C.

At that, the thrust bearing design shall not allow overheating, vibration, oil spraying and getting into the ventilation duct at all operation modes, including start-ups and stops, as well as operation at maximum runaway rotation speed.

Designing of the bearings with a two-row arrangement of the segments, that allowed to maintain the level of the specific load without increasing the size of the segments. In the two-row design of the thrust bearing (two rows of chambers) the load through the hub and disc is transferred to segments located in two circular rows. The segments are laid on the supporting plates, which are supported by the spherical surfaces of the supporting bolts screwed into the balancing beam. The balancer is on a cylindrical support. Segments are connected in pairs by means of the balancing beam. The load between the external and internal segments in the pair is automatically distributed according to the law of the double-sided lever (the first-order lever).

Its disk provides electrical isolation from the housing to prevent leakage of the bearing currents. Insulation monitoring is performed on a stopped Hydrogenerator-motor without disassembling of the bath. The segments of the thrust bearing are made with elastic metal-plastic straps (with fluoroplastic coating).

At that the specific pressure on to the thrust bearing segments are limited with 6 MPa. The design provides uniform distribution of the load on to the segments with tolerance of ±10%.

The thrust bearing is arranged in the oil bath protected from water ingress and is self-lubricating. For its lubrication oil of Grade Tp-30 is used in accordance with GOST 9972-74 (or its analogue). It is cooled with water circulating through the oil coolers arranged in the oil bath to temperatures corresponding to the established norms.

The thrust bearing ensures reliable operation in both directions of rotation of the Hydroaggregate as well as the start of the Hydrogenerator after its long term stop without usage of special measures.

2. Determining of Maximum Admissible Operation Time of the Thrust Bearing Unit

The classical approach to calculating of the fatigue state of the thrust bearing unit reduces to taking into account only uniform distribution of the load (tension-compression) over the entire area of the segments, the diagram of which is shown in Figure 4.

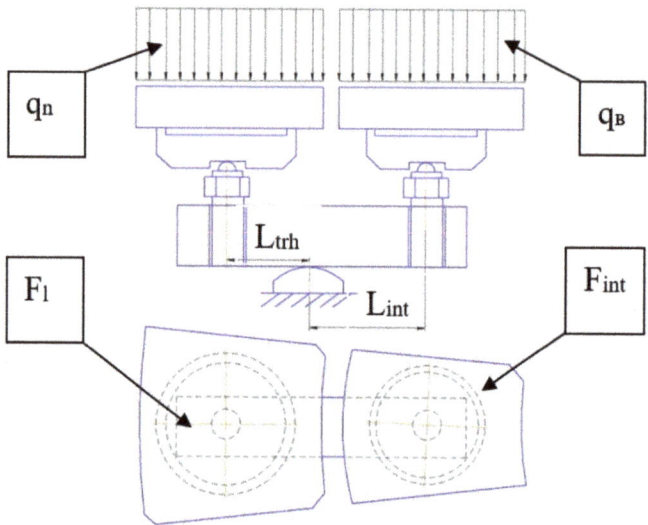

Figure 4. Loading Diagram of the double-row thrust bearing segments.

The condition for the equality of the moments of forces acting on the segments shall be written in the following form:

$$P_{ext}l_{ext} = P_{int}l_{int} \text{ or } q_{ext}F_{ext}l_{ext} = q_{int}F_{int}l_{int}, \qquad (1)$$

where P_{ext} and P_{int}—are the loads on to the external and internal segments; q_{ext} and q_{int}—are specific loads on to the working surface of the external and internal segments; F_{ext} and F_{int}—are the areas of the working surface of the external and internal segments; l_{ext} and l_{int}—the lengths of the arms of the lever of the external and internal segments.

In the Commonwealth of Independent States states more than 70 double-row thrust bearings were manufactured and installed on the aggregates of various hydroelectric power plants. They are designed for the total load from 20 up to 35 MN and specific load from 3.5 up to 4.2 MPa. The average circumferential speed in the internal row of segments comprises of 8.0–24.0 m/s, and on the external one comprises of 10.8–31.0 m/s.

When carrying out the analytical calculation, the rigidity of the entire design is searched for by adding of the chamber rigidity due to the compressibility of oil and the rigidity of the empty chamber:

$$C = 0.85A + B, \qquad (2)$$

where A—the rigidity of the chamber due to the compressibility of oil; B—the rigidity of the empty chamber.

The submitted approach can be correct in the first approximation and used at the stage of preliminary outline design.

However, for the working project the calculation diagram shall be radically revised. The operation peculiarities of the chambers of the near and the far rows are different therefore it is necessary to take into account the entire three-dimensional figure of the loads effect on the thrust bearing that is possible to carry out only at solid state modeling of the entire design in a three-dimensional formulation.

Let us consider the three-dimensional calculation design model of the thrust bearing chamber, made by the finite element method (FEM) (SolidWorks Simulation). The design diagram and the limiting conditions of the thrust bearing unit loading under the action of the total axial load from water pressure and rotating parts of the pump-turbine unit, as well as the mass of the rotating parts of the pump-turbine and the water reaction are submitted in Figure 5.

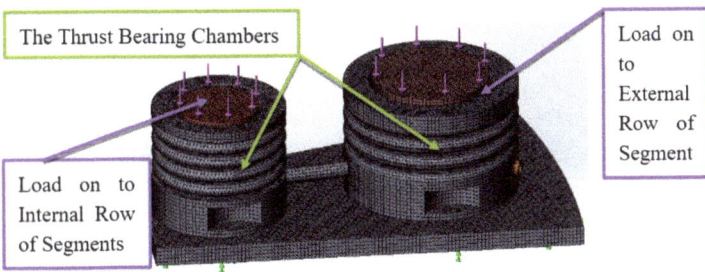

Figure 5. Calculation Diagram and limiting conditions of the thrust bearing unit loading.

As it can be seen from Figure 6 the design vessel (the thrust bearing chamber) is closed. It means that the physical properties of oil shall be the same everywhere and, consequently, pressure in all points of the chamber shall be constant ($P = const$).

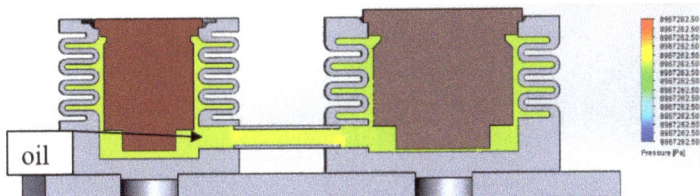

Figure 6. Diagrams of pressure inside the thrust bearing chambers.

The successive loading of the thrust bearing was considered at small increments of the load that allowed observing insignificant change in the oil elasticity modulus for each iteration. At that the rigidity of the thrust bearing changes significantly at all stages of operation.

The diagrams of pressure inside the thrust bearing chambers are submitted in Figure 6. The calculations are carried out with the help of the package SolidWorks Flow Simulation. As it can be seen from Figure 6 the values of pressures are practically constant.

When carrying out the fatigue calculation, the following factors shall be taken into account:
- technological factor;
- geometrical factor;
- the surface roughness factor (taking into account additional influence of roughness to local stresses and, consequently, on the fatigue strength of the component);
- the influence factor of surface hardening (taking into account influence (residual stress, hardness) of the changed state of the surface to fatigue strength of the corresponding technological procedure)

Thermally treated steel 35 GOST 1050-2013 shall be considered as the basic material for the chambers of the thrust bearing unit. As its foreign analogues the brands submitted in scientific work [20] can be considered. Check for compliance of all the mechanical properties of the material of the thrust bearing chamber shall be in accordance with GOST 8479-70 for Group V. The fatigue curve equation under rigid loading was performed in double logarithmic coordinates according to GOST 25.507-85.

The fatigue calculation algorithm assumes the choice of the maximum effective load from hydraulic shock and the masses of the rotating parts of the hydrogenerating unit. In this case, a load is set that varies in time depending on the of rotational speed of the hydrogenerating unit. In this case, the full cycle is considered to be full-loading followed by unloading. The maximum stress at which fatigue failure occurs after a given number of load cycles. The permissible number of load cycles is determined based on the regulatory documents for hydrogenerating units (the maximum number of starts). To calculate the compliance of the supports, a correlation of oil elasticity is put depending on the load.

As per the results of the research it can be seen (see Figure 7) that run out (wearing out) of the design should occur much later than the designed operation life of the design. Percentage of damage.

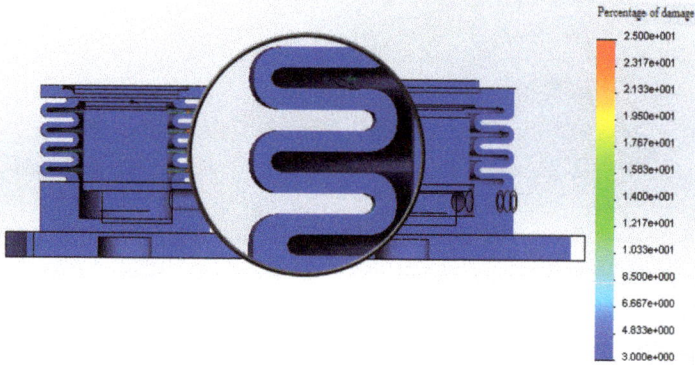

Figure 7. Percentage of damage of the thrust bearing chamber.

3. Discussion

All over the world, there are 2 types of the thrust bearings: on rigid and hydraulic supports. All existing algorithms are designed for rigid supports, because they are most often used in Hydrogenerators with medium power. For European schools of mechanical calculation, the main trend was a decrease in weight and overall dimensions indices due to an increase in mechanical stresses approaching the allowable ones. This algorithm was normally transferred to body parts and individual parts of the rotor group. However, due to the fact that hydraulic thrust bearings are used on hydro generators of maximum power for these types of bearings, this algorithm has not been fully implemented. Creation in Ukraine of Hydrogenerators with power over 300 MW for the first time made it possible to switch to accurate calculations by the FEM method of thrust bearings, taking into account the properties of the liquid inside the support. In our opinion, Chinese concerns will continue to use similar algorithms for the transition to high power Hydroelectric Units.

4. Conclusions

In submitted work, the stressed state of the thrust bearing unit of Hydrogenerator of the limiting power is considered. The main disadvantages of the existing designs are indicated, as well as the most dangerous technical damage. To ensure reliable operation of the Hydroaggregate, a methodology for performing fatigue calculation is proposed, which allows to take into account technological, operational and constructive factors, by FEM modeling of the design. It is shown that the design under consideration satisfies the requirements of GOST 14965-80, and the required service life for 40 years is reliably ensured.

Author Contributions: Writing—Review and editing O.T., D.K., I.K., M.A., V.N. All authors have read and agreed to the published version of the manuscript.

Funding: This research received no external funding.

Institutional Review Board Statement: Not applicable.

Data Availability Statement: Not applicable.

Conflicts of Interest: The authors declare no conflict of interest.

References

1. Abramov, A.I.; Ivanov-Smolenskyi, A.V. *Designing of Hydrogenerators and Synchronous Compensators: Aducational Aid for Technical Colleges*; Higher School: Moscow, Russia, 1978; p. 312.
2. Bovsunovskyi, A.P.; Chernousenko, O.Y.; Shtefan, Y.V.; Bashta, D.A. Fatigue damage and destruction of rotors of steam turbines as a result of torsional oscillations. *Strength Probl.* **2010**, *1*, 144–151.
3. DeCamillo, S.M.; Dadouche, A.; Fillon, M. Thrust Bearings in Power Generation. In *Encyclopedia of Tribology*; Wang, Q.J., Chung, Y.W., Eds.; Springer: Boston, MA, USA, 2013. [CrossRef]
4. Selak, L.; Butala, P.; Sluga, A. Condition monitoring and fault diagnostics for hydropower plants. *Comput. Ind.* **2014**, *65*, 924–936. [CrossRef]
5. Liu, X.; Luo, Y.; Wang, Z. A review on fatigue damage mechanism in hydro turbines. *Renew. Sustain. Energy Rev.* **2016**, *54*, 1–14. [CrossRef]
6. Iliev, H. Failure Analysis of Hydro-Generator Thrust Bearing. *Wear* **1999**, *225–229*, 913–917. [CrossRef]
7. Xu, W.; Yang, J. The Air Lubrication Behavior of a Kingsbury Thrust Bearing Demonstration. *Hindawi Int. J. Rotating Mach.* **2021**, *2021*, 6690479. [CrossRef]
8. Nowicki, R.; Morozov, N. Thrust Bearing Monitoring of Vertical Hydro-Turbine-Generators. In Proceedings of the June 2017 Conference: 1st World Congress on Condition Monitoring, London, UK, 13–16 June 2017; Volume 174, p. 15.
9. Simmons, J.E.L.; Advani, S.D. Paper II(iii) Michell and the development of tilting pad bearings. In *Tribology Series*; Elsevier: Amsterdam, The Netherlands, 1987; Volume 11, pp. 49–56.
10. Simmons, J.E.L.; Knox, R.T.; Moss, W.O. The development of PTFE (polytetrafluoroethylene)-faced hydrodynamic thrust bearings for hydrogenerator application in the United Kingdom. *Proc. Inst. Mech. Eng. Part J J. Eng. Tribol.* **1998**, *212*, 345–352. [CrossRef]
11. Yang, S.; Zheng, W.; Jiang, M.; Pei, S.; Xu, H. A comparative experimental study on large size center and bi-directional offset spring-bed thrust bearing. *Proc. Inst. Mech. Eng. Part J J. Eng. Tribol.* **2020**, *234*, 134–144. [CrossRef]
12. Kingsbury, I. *A General Guide to the Principles, Operation and Troubleshooting of Hydrodynamic Bearings*; Kingsbury Inc.: London, UK, 1997.
13. Dabrowski, L.; Wasilczuk, M. A Method of Friction Torque Measurement for a Hydrodynamic Thrust Bearing. *J. Tribol.* **1995**, *117*, 674–678. [CrossRef]
14. Fillon, M.; Wodtke, M.; Wasilczuk, M. Effect of presence of lifting pocket on the THD performance of a large tilting-pad thrust bearing. *Friction* **2015**, *3*, 266–274. [CrossRef]
15. Fillon, M. Thermal and Deformation Effects on Tilting-Pad Thrust and Journal Bearing Performance. In Proceedings of the III Congresso Ibérico de Tribologia, IBERTRIB, Guimarães, Portugal, 16–17 June 2005.
16. Simmons, J.E.L.; Advani, S.D. Michell and the development of tilting pad bearings. In *Fluid Film Lubrication–Osborne Reynolds Centenary*; Elsevier: Amsterdam, The Netherlands, 1987.
17. Kingsbury, A. Thrust Bearing. U.S. Patent 1102276, 7 July 1914.
18. Gruboi, A.P.; Diakov, V.I.; Kubanov, V.G.; Saltovskaya, D.A. Increase of reliability and efficiency of temperature control of the bearings of vertical Hydroaggregates. *Hydroenergetics Ukr.* **2014**, *2–3*, 25–27.
19. Vakulenko, A.N.; Kobzar, K.A.; Tretiak, A.V.; Gakal, P.G. Recognition of emergency situations of large Hydrogenerators (Hydrogenerators-Motors) by multifactor analysis of complex stressed condition of units and parts. *Hydroenergetics Ukr.* **2015**, *1–2*, 23–27.
20. Troschenko, V.T.; Sosnovskyi, L.A. *Metals and Alloys Fatigue Resistance: Reference*; Naukova Dumka: Kiev, Ukraine, 1987; p. 347.

Article

Mathematical Models and Nonlinear Optimization in Continuous Maximum Coverage Location Problem

Sergiy Yakovlev [1,2,*], Oleksii Kartashov [1] and Dmytro Podzeha [1]

[1] Mathematical Modelling and Artificial Intelligence Department, National Aerospace University "Kharkiv Aviation Institute", 61072 Kharkiv, Ukraine; alexeykartashov@gmail.com (O.K.); dima.podzega@gmail.com (D.P.)
[2] Institute of Information Technology, Lodz University of Technology, 90-924 Lodz, Poland
* Correspondence: s.yakovlev@khai.edu

Abstract: This paper considers the maximum coverage location problem (MCLP) in a continuous formulation. It is assumed that the coverage domain and the family of geometric objects of arbitrary shape are specified. It is necessary to find such a location of geometric objects to cover the greatest possible amount of the domain. A mathematical model of MCLP is proposed in the form of an unconstrained nonlinear optimization problem. Python computational geometry packages were used to calculate the area of partial coverage domain. Many experiments were carried out which made it possible to describe the statistical dependence of the area calculation time of coverage domain on the number of covering objects. To obtain a local solution, the BFGS method with first-order differences was used. An approach to the numerical estimation of the objective function gradient is proposed, which significantly reduces computational costs, which is confirmed experimentally. The proposed approach is shown to solve the maximum coverage problem of a rectangular area by a family of ellipses.

Keywords: geometric object; coverage; optimization; computational geometry software; Python

1. Introduction

The maximum coverage problem is a classical problem in computer science, computational complexity theory, and operations research. In the original formulation, the problem is as follows.

There are a family of sets $\Omega = \{S_1, \ldots, S_n\}$ and integer number $1 \leq k \leq n$. Sets may have some elements in common. The objective is to find such a family $\tilde{\Omega} \subseteq \Omega$ of k sets from Ω so that the maximum number of elements is covered, i.e., the union $\bigcup_{S_i \subseteq \tilde{\Omega}} S_i$ of the selected sets has the maximum size.

The great interest of scientists in the problem of maximum coverage is associated with a wide range of its practical applications. The concept of coverage is associated with video surveillance cameras, emergency warning systems, mobile communications and the Internet, radio stations, environmental sensors, service systems, etc. This leads to the need to cover certain areas and is associated with an analysis of the location of objects of various nature. A major category of location analysis approaches involves coverage when an entity provides services based on spatial proximity. This may be based on travel time, line-of-sight, or hearing factors. These spatial footprints reflect services and may be regular or irregular, contiguous, or fragmented in area. One of the most common approaches to location modeling is the maximum coverage location problem (MCLP). Such a problem is the subject of study for both operations research theory and computational geometry. In this case, the goal is to place several objects in a given region in such a way as to maximize (minimize) some objective function associated with the considered type of objects.

In most of these problems, a demand point is considered to be covered by a facility if the distance or travel time between them is less than a given predetermined value. It is desirable that all points of demand are served by at least one enterprise. However, often full coverage of demand in the region is impossible due to restrictions on the possible number of objects to be placed. In this case, the task is to cover the maximum possible area of demand through the efficient use of limited resources.

Note that the problems associated with determining the location of objects are both discrete and continuous in nature. In the discrete case, there is a finite set of possible locations of potential objects, and it is required to choose their optimal location. Such tasks arise in the placement of physical services, when it is possible to predetermine a finite set of admissible locations. At the same time, there is a wide class of problems when it is allowed to place objects in the entire space or its continual subdomain. Such tasks are continuous (for example, in telecommunications networks, when placing sensors or radar stations, etc.). In this case, continuity conditions can be imposed not only on the sets of admissible locations, but also on the service area. This article is devoted to mathematical modeling and solving these coverage optimization problems.

2. Background

Reviews [1–5] are devoted to the issues of object location analysis based on the modeling of coverage problems. The main approaches to location analysis consider coverage in the sense that an object provides services within spatial proximity, i.e., each object has a spatial footprint that defines a service area. Article [6] provides extended overview of the maximal covering location problem, highlighting the use, application, solution, evolution, and generalization of this important location analytic approach.

The problem of placing a certain number of objects within an acceptable service distance to maximize demand in a given region was posed and formalized by Church and ReVelle [7]. This problem is called the maximum coverage location problem (MCLP). To solve this problem, two sets of discrete locations were considered, representing demand and possible locations. As a result, the MCLP is formulated as an integer linear programming problem that can be solved in various ways. Variations arise due to the use of different methods for determining the area of coverage of a potential object or assigning weights to sites on several factors and parameters.

The standard form of MCLP considers a finite set of potential locations for objects, and a set of discrete points represents demand. The paper [8] proposes a generalization that allows both facilities and demand points to be placed continuously on a plane. Such a problem is called planar MCLP. An exact planar MCLP solution, assuming that the need for coverage exists anywhere in the surrounding area and that objects can also be arbitrarily placed in this area, is impossible due to the NP-hard nature of the problem [9]. Several heuristic methods have been proposed in an attempt to obtain acceptable solutions. For example, paper [9] describes a greedy algorithm for MCLP that selects at each step the location containing the largest number of uncovered elements. This greedy algorithm is claimed to be the best approximation algorithm in polynomial time.

In the work [10], a planar MCLP with partial coverage and rectangular demand and service zones is studied. The problem is to position a given number of rectangular service zones on the two-dimensional plane to partially cover a set of existing (possibly overlapping) rectangular demand zones such that the total covered demand is maximized. Based on the properties of the model, the possibility of a significant reduction in the search space is theoretically substantiated and a modification of the branch and bound algorithm for solving the problem is presented. The authors of [10,11] use a computational geometry approach to study partial coverage with rectangular and circular service areas. However, according to [12], optimality is guaranteed only in some special cases.

The paper [13] presents a fuzzy maximally covering location problem in which travel time between any pair of nodes is considered to be a fuzzy variable. A hybrid algorithm of fuzzy simulation and simulated annealing is used to solve MCLP. Some numerical examples are solved and analyzed to show the performance of the proposed algorithm.

In the article [14], a mixed integer non-linear programming model was formulated for distributing the position of a number of rectangular objects of unequal area in a continuum of a planar plant site with a predetermined fixed area. A continuous approach to the problem is taken. Constraints are developed to eliminate the possible overlap between the different facilities. A novel simulated annealing algorithm is developed to solve large instances of the problem. A unique heuristic algorithm is used for initial solution.

We especially note the work [15], in which the continuous setting of MCLP is analyzed. Continuity conditions are imposed on the location of objects, however, the discrete structure of the service domain is preserved, taking into account the graph of relations between enterprises. The problem of mixed integer non-linear programming is formulated. At the same time, taking into account the geometric properties, the authors propose its equivalent reformulation as an integer linear programming problem.

Other approaches use the division of demand into several smaller regions and the ratio between fully covered or uncovered areas [8,16–18]. A facility is assumed to cover the demand for a region if it covers the entire region. Partially covered areas are considered as uncovered areas. Such coverage is called binary. Consideration of the binary coverage makes it possible to obtain a finite set of potential service points determined by the set of polygon intersection points. As a result, an integer linear programming problem is formulated. However, although binary coverage makes planar MCLP manageable, this approach is prone to significant errors because all partially covered regions, which may have a significant coverage area, are left out and ignored in the final solution [12].

Note that obtaining even a partial maximum coverage is not trivial. Therefore, as an auxiliary problem, one often considers partial coverage by only one object. In the paper [19], a method for solving the problem of determining the location of a circular service area in a polygonal zone is described based on the calculation of the mean axis of the zone. Taking into account partial coverage, in the work [20] it is proposed to approximate the optimal position of a single circular footprint with bounded error in an unconnected area with holes bounded by line or circle segments. The authors describe an efficient algorithm for accurately calculating the overlap area of a circle and a polygon, and introduce the concept of a coverage map for more information. The paper [21] presents improved algorithms for matching two polygonal shapes to approximate their maximum overlap. The run time of algorithms is studied.

The article [22] describes an approach to solving the problem of locating a disk so that its overlap area with a piecewise circular domain is near-optimal when considering partial coverage. The concept of an overlap area map is introduced. Parallel algorithm on graphics processing units are discussed for calculating the overlap area map and deriving a set of near-optimal locations from the overlap area map. Particular attention is paid to the process of calculating the overlap area. The authors visualize the information obtained, highlighting the optimal location, and analyze the experimental results obtained by implementing the proposed algorithms.

The study [23] presents an approach to the maximum coverage of a polygonal area by disk services of a given radius. Of particular interest is a parallel optimization method using simulated annealing based on a perturbation strategy and a probabilistic estimate of the covered area of a polygon. The system provides in a reasonable run time fairly good location of disks, starting with an arbitrary initial solution. The paper [24] considers the problem of covering compact polygonal set by identical circles of minimal radius. A mathematical model of the problem based on Voronoi polygons is constructed and its characteristics are investigated. A modification of the Zoutendijk feasible directions method is developed to search local minima. A special approach is suggested to choose starting points. Many computational examples are given.

Access to and solution of the MCLP is also possible in *GIS* software packages. For example, structuring and solving an MCLP application instance in *ArcGIS* and *TransCAD*, among others, is easily accomplished. In *ArcGIS*, there are location-allocation functions in network analyst, and, in particular, "maximize coverage" would enable an MCLP to be set up as the problem type. Similarly, in *TransCAD*, one would use "facility location analysis" to accomplish this. There are in fact a number of reported application studies that have relied on this basic approach using *GIS*, such as [25,26]. It is important to note that *GIS* packages facilitating access to the MCLP typically solve problems using a heuristic [4].

The evolution of the application of MCLP has occurred concurrently in concert with the development, growth, and maturation of *GIS*. The MCLP access and solution is possible in various *GIS* software packages. The authors of [3,27] discuss that *GIS* is a combination of hardware, software, and procedures enabling data management and spatial query. Key components of *GIS* include data capture, modeling, management, manipulation, analysis, and display. It is important that in addition to generating and working with spatial information, *GIS* supports a number of analytical functions, many of which facilitate the MCLP specification and application. In the paper [4], the capabilities of a *GIS* for carrying out suitability analysis, as well as operations associated with containment, overlay, distance, buffering, and spatial interpolation, among others. Much spatial information now exists, expediting coverage location analysis as well.

3. Materials and Methods

3.1. Mathematical Model of Continuous Maximum Coverage Location Problem

Let us formulate the continuous MCLP as a geometric problem. In the space R^d, $d = 2, 3$ there are a certain domain S_0 and a family of geometric objects $\{S_1, \ldots, S_n\}$ of a given shapes and sizes. We will say that a family of objects $\{S_1, \ldots, S_n\}$ covers some subdomain $\tilde{S} \subseteq S_0$ if each point $\tilde{S}$ belongs to at least one of the objects of the family. The problem is to find such a location of objects $S_1, \ldots, S_n$ that they cover as much of the domain S_0 as possible. S_0 is called the coverage domain, and $S_1, \ldots, S_n$—the covering objects.

To formalize the continuous MCLP we make the following notations: $J_n = \{1, \ldots, n\}$, $J_n^0 = J_n \cup \{0\} = \{0, 1, \ldots, n\}$. Geometric object $S_i \subset R^d$, $i \in J_n^0$ means a geometric set of points $P \in R^d$ that satisfy inequalities $f_i(P, \mathbf{m}^i) \geq 0$. The equation $f_i(P, \mathbf{m}^i) = 0$ defines the boundary of the object S_i and determines its shape. In the general case, functions $f_i(P, \mathbf{m}^i)$, $i \in J_n^0$ contain constants $\mathbf{m}^i = (m_1^i, \ldots, m_{\alpha_i}^i)$, which are called metric parameters of the shape of the object S_i. These parameters determine the linear sizes of the corresponding objects.

Consider a fixed Cartesian coordinate system in space R^d, $d = 2, 3$, and with each object $S_i \subset R^d$, $i \in J_n^0$, connects own coordinate system, the beginning of which is called the pole of this object. The relative position of these system will be characterized by parameters $\mathbf{p}^i = (p_1^i, \ldots, p_{\beta_i}^i) = (\mathbf{v}^i, \theta^i)$, $i \in J_n^0$, where $\mathbf{v}^i$—a coordinate vector of the pole of the object S_i in a fixed coordinate system, and θ^i—a vector of angular parameters that determine the relative position of the axes of own and fixed coordinate systems. The coordinates of the vector $\mathbf{p}^i = (p_1^i, \ldots, p_{\beta_i}^i)$ indicates the position of the object S_i in R^d, $d = 2, 3$ and are called placement parameters. In this case, the position of the object S_i relative to the fixed coordinate system is given by the equation of the general position, which has the form

$$F_i(P, \mathbf{m}^i, \mathbf{v}^i, \theta^i) = f_i\left(A(p - \mathbf{v}^i), \mathbf{m}^i\right) = 0, \qquad (1)$$

where A is the orthogonal operator expressed through angular parameters θ^i.

In the papers [28,29], methods for constructing configurations of geometric objects are described, and classes of packing, layout, and covering configurations are identified. A configuration space of geometric objects is constructed, the generalized variables of which are placement parameters and metric parameters of objects.

Let us apply these results to build a model of the continuous MCLP. Let $\Omega = \{S_1, \ldots, S_n\}$ be a family of geometric objects with metric parameters of the shape $\mathbf{m}^i = (m_1^i, \ldots, m_{\alpha_i}^i)$ and placement parameters $\mathbf{p}^i = (p_1^i, \ldots, p_{\beta_i}^i) = (\mathbf{v}^i, \theta^i)$, $i \in J_n^0$. A parameters $\mathbf{g}^i = (\mathbf{m}^i, \mathbf{p}^i) = (m^i_1, \ldots, m^i_{\alpha_i}, p^i_1, \ldots, p^i_{\beta_i})$ is called a generalized variable of the object S_i. The dimension of the vector $\mathbf{g}^i$ is equal to $\gamma_i = \alpha_i + \beta_i$. The object S_i with the generalized variables $\mathbf{g}^i$ denote by $S_i(\mathbf{g}^i)$, $i \in J_n^0$.

Note that, depending on the specifics of the problem, some generalized variables can be fixed. To indicate this we will use a cap over the corresponding variable. In continuous MCLP, placement parameters of the coverage domain S_0 are assumed to be fixed and usually equal to $\hat{\mathbf{p}}^0 = (0, \ldots, 0)$. Metric parameters $\mathbf{m}^0$ in the general case are variable, unless otherwise specified by the problem. Considering problems when the sizes of S_0 are known, we will assume $\mathbf{m}^0 = \hat{\mathbf{m}}^0$, so $\hat{\mathbf{g}}^0 = (\hat{\mathbf{m}}^0, \hat{\mathbf{p}}^0)$.

Using set-theoretic operations, we build a complex object

$$\Omega = S_0 \cap \bigcup_{i=1}^n S_i. \tag{2}$$

We assume that Formula (2) determines the structure of a complex object Ω, i.e., the rule of its formation.

Having determined the generalized variables $\mathbf{g}^i$, $i \in J_n^0$ of the constituent objects S_i, we form a parameterized object

$$\Omega(\mathbf{g}) = \Omega(\hat{\mathbf{g}}^0, \mathbf{g}^1, \ldots, \mathbf{g}^n) = S_0(\hat{\mathbf{g}}^0) \cap \bigcup_{i \in J_n} S_i(\mathbf{g}^i). \tag{3}$$

In $\mathbb{R}^d$, $d = 2, 3$, each fixed vector $\hat{\mathbf{g}} = (\hat{\mathbf{g}}^0, \hat{\mathbf{g}}^1, \ldots, \hat{\mathbf{g}}^n)$ of generalized variables corresponds to a complex geometric object of a given structure

$$\Omega(\hat{\mathbf{g}}) = S_0(\hat{\mathbf{g}}^0) \cap \bigcup_{i \in J_n} S_i(\hat{\mathbf{g}}^i). \tag{4}$$

Depending on the dimension of $\mathbb{R}^d$, $d = 2, 3$, we calculate the measure (area or volume) of the formed object $\Omega(\hat{\mathbf{g}})$, which we denote by $\mu(\Omega(\hat{\mathbf{g}}))$. Thus, an arbitrary set of generalized variables $\mathbf{g} = (\hat{\mathbf{g}}^0, \mathbf{g}^1, \ldots, \mathbf{g}^n)$ can be associated with a measure of a complex object $\Omega(\mathbf{g})$ of a given structure, i.e., define a function of generalized variables

$$\omega_\Omega(\mathbf{g}) = \mu(\Omega(\hat{\mathbf{g}})). \tag{5}$$

The described class of functions is called ω-functions [30].

For the formalization of the ω-function, we introduce the characteristic function

$$\lambda_\Omega(P, \mathbf{g}) = \begin{cases} 1, & \text{if } P \in \Omega(\mathbf{g}); \\ 0, & \text{if } P \notin \Omega(\mathbf{g}). \end{cases} \tag{6}$$

If $P \in \mathbb{R}^2$, then

$$\omega_\Omega(\mathbf{g}) = \iint \lambda_\Omega(P, \mathbf{g}) dP \tag{7}$$

and for $P \in \mathbb{R}^3$

$$\omega_\Omega(\mathbf{g}) = \iiint \lambda_\Omega(P, \mathbf{g}) dP. \tag{8}$$

Properties of ω-functions were considered in the article [31].

Using ω-functions allows us to state the maximum coverage problem as the following nonlinear unconstrained optimization problem

$$\omega_\Omega(\mathbf{g}) \to \max \tag{9}$$

where ω-function is defined by the Formula (5), and the structure Ω is given as (2).

This statement of the continuous MCLP follows from the fact that the function $\omega_\Omega(\mathbf{g})$ determines the dependence of the measure (area) of the partial coverage of the domain on the generalized variables **g** of the family of covering objects. It is the maximization of this measure that is the objective of the problem.

Thus, the solution of the continuous MCLP is inextricably linked, on the one hand, with methods for calculating the function $\omega_\Omega(\mathbf{g})$, and, on the other hand, with the choice of an effective method for solving the optimization problem (9).

Therefore, the general scheme for solving the problem includes the following stages:

➢ Formation of initial data on the coverage domain and the family of covering objects;
➢ Choice of generalized coverage configuration variables;
➢ Calculating a measure of the covered part of the coverage area;
➢ Choice of local optimization method for solving problem (9);
➢ Estimation of the gradient of the objective function, taking into account the geometric properties of the problem;
➢ Choice of global optimization method.

The first two stages were discussed earlier. The next stages are described in the next subsection.

3.2. Computer Geometry and Optimization Software

To formalize the function, one can use the analytical approach [31], which consists of constructing an equation for the boundary of a complex geometric object based on the known equations for the boundaries of composite objects. However, this approach is very time consuming and requires large computational costs, especially considering that the objects depend on parameters (generalized variables). The papers [32,33] propose ways to formalize the coverage conditions using *phi*-functions, but only for a narrow class of covering objects, such as rectangles, circles, and spheroids [24,34,35].

At the same time, considering the issue of working with geometric objects of a given shape, you can find a powerful list of libraries that allow you to work with such figures, in particular, *SymPy, Shapely, CGAL, SpaceFuncs*, and many others. Based on the analysis of existing libraries, taking into account the necessary functionality to calculate the ω-function of spatial configurations (complex geometric objects), we chose the *Shapely* library [36]. *Shapely* is a *Python* package for theoretical analysis of points set and plane object manipulation using *Python ctypes* module functions from the well-known *GEOS* library. Note that *GEOS* as a port of the *Java Topology Suite (JTS)* is the geometry mechanism of the *PostGIS* spatial extension for the *PostgreSQL* database. On the one hand, the *Shapely* package is deeply rooted in the conventions of the world of geographic information systems *(GIS)*. On the other hand, this package seeks to be equally useful for programmers working on non-traditional issues, including image processing and geometric design. It is with the help of the *Shapely* library that it is possible to construct complex shapes through set-theoretic operations on geometric objects (union, intersection, symmetric difference, product) from a set of basic shapes (circle, ellipse, polygon, etc.), as well as perform the same operations on arbitrary complex figures, built with the help of basic. Using *the unary_union(·)* function in *shapely.ops* allows you to build unions at the same time, which is much more efficient than sequential accumulation using the *union(·)* operation. The set of affine transformation functions is contained in the *shapely.affinity* module, which transforms geometric shapes by directly assigning coefficients to the affine transformation matrix, or by means of a specially named transformation (rotation, scale, etc.). Note that geometric objects are created in the typical *Python* way, using the classes themselves as instance factories.

The most important feature of the *Shapely* library is the ability to calculate the area of any complex object using the *area* field.

The use of modern computer geometry software allows us to offer new effective methods for solving problems of maximum coverage. In particular, in *Python* has developed a package of applied mathematical procedures *SciPy* based on the *Numpy Python* extension. With *SciPy*, the *Python* interactive session becomes the same full-fledged data processing and prototyping environment for complex systems, such as any versions of MATLAB, IDL, Octave, R-Lab, and SciLab later than 2015.

On the other hand, the properties of the MCLP make it possible to use gradient methods of local optimization using first order differences. To do this, consider the k-th component of the vector $\hat{g}^i = (\hat{g}^i_1, \ldots, \hat{g}^i_k, \ldots, \hat{g}^i_{\gamma_i})$ corresponding to the object S_i and give it an increment δ. Denote

$$\widetilde{g} = (\hat{g}^0, \ldots, \hat{g}^{i-1}, \hat{g}^i_1, \ldots, \hat{g}^i_k + \delta, \ldots, \hat{g}^i_{\gamma_i}, \hat{g}^{i+1}, \ldots, \hat{g}^n). \tag{10}$$

Then, to estimate the increment of a function $\omega_\Omega(\hat{g})$ at a point $\hat{g} = (\hat{g}^0, \ldots, \hat{g}^{i-1}, \hat{g}^i_1, \ldots, \hat{g}^i_k, \ldots, \hat{g}^i_{\gamma_i}, \hat{g}^{i+1}, \ldots, \hat{g}^n)$, we obtain

$$\Delta \omega_\Omega(\hat{g}) = \mu(\Omega(\widetilde{g})) - \mu(\Omega(\widetilde{g})), \ i \in J_n, \tag{11}$$

where the vectors $\hat{g}$ and $\widetilde{g}$ are differ in only one coordinate $\hat{g}^i_k$ changed to δ.

This approach to determining the increment of a function $\omega_\Omega(\hat{g})$ is universal, but it requires the calculation of the values of function at points $\hat{g}$ and $\widetilde{g}$.

Using the properties of ω-functions, one can propose an approach that has less computational complexity.

Let us form a symmetric matrix of areas of pairwise intersections of objects

$$M = [\mu_{ij}]_{(n+1) \times (n+1)'} \tag{12}$$

where

$$\mu_{ij} = \mu\left(S_i(\hat{g}^i) \cap S_j(\hat{g}^j)\right), \ i,j \in J^0_n. \tag{13}$$

As a rule, such a matrix M is highly sparse. Let us use this property.

For the k-th component of the vector $\hat{g}^i = (\hat{g}^i_1, \ldots, \hat{g}^i_k, \ldots, \hat{g}^i_{\gamma_i})$, corresponding to the object S_i, we will increment δ and form the object $S^k_i(\widetilde{g})$, where $\widetilde{g}$ has the form (10). Then, the increment of the function $\omega_\Omega(g)$ at the point $\hat{g} = (\hat{g}^0, \ldots, \hat{g}^n)$ will be equal to

$$\Delta^i_k \omega_\Omega(\hat{g}) = \sum_{j \in J^0_n} \left(\mu\left(S^k_i(\widetilde{g}) \cap S_j(\hat{g}^j)\right) - \mu_{ij} \right) \tag{14}$$

where the summation is over all $j \in J^0_n$, such that $\mu_{ij} \neq 0$.

The experiments described in the next section confirm that it takes an order of magnitude less time to evaluate the increment of a function using Formula (14) than when using Formula (11).

4. Results

To choose and justify the method for solving the optimization problem (9), we studied the dependence of computational resources on the dimension of the problem, i.e., number of covering objects. We conducted a series of experiments during which we evaluated:

- Time for calculating the area of a complex object, the structure of which is determined by Formula (4);
- Run time of local optimization for an arbitrary starting point;
- Run time of forming the area matrix of pairwise intersections of objects.

Computational experiments were carried out using a computer with the following configuration: Intel Core i7-5557U processor, CPU Speed 3.1 GHz, 2 cores, 4 threads; RAM 16 GB DDR3 1866 MHz; Graphics processor Intel Iris Graphics 6100 with 1.5 GB of video memory; SSD 512 GB; and operating system Mac OS X11.0 Big Sur.

The calculations were made using *Python* packages, such as *Shapely* and *SciPy*.

We chose a rectangle as the coverage domain S_0, and ellipses as the covering objects S_i, $i \in J_n$. The metric parameters $\hat{m}^0 = (a_0, b_0)$ of S_0 are lengths of their sides, and the placement parameters are $\hat{p}^0 = (0,0,0)$. The metric parameters $\mathbf{m}^i = (a_i, b_i)$, $i \in J_n$ of ellipses are their semi-axis, and the placement parameters $\mathbf{p}^i = (x_i, y_i, \theta_i)$, $i \in J_n$ are the coordinates (x^i, y^i) of the symmetry centers and the angle of rotation θ^i. The general equation of the ellipse in accordance with (1) has the form

$$1 - \frac{((x - x_i) \cos \theta_i + (y - y_i) \sin \theta_i)^2}{a_i^2} + \frac{(-(x - x_i) \sin \theta_i + (y - y_i) \cos \theta_i)^2}{b_i^2} = 0. \quad (15)$$

So, $\hat{g}^0 = (\hat{m}^0, \hat{p}^0) = (a_0, b_0, 0, 0, 0)$, $g^i = (\hat{m}^i, \mathbf{p}^i) = (a_i, b_i, x_i, y_i, \theta_i)$, $i \in J_n$, $g = (\hat{g}^0, g^1, \ldots, g^n) = (a_0, b_0, 0, 0, 0, a_1, b_1, x_1, y_1, \theta_1, \ldots, a_n, b_n, x_n, y_n, \theta_n)$.

For each n from 10 to 500 with step 10, we will randomly form the initial data of the problem, generating semi-axes (a_i, b_i), $i \in J_n$ of ellipses evenly distributed on the interval $(0, 100)$. Then, we calculate the total area of the constructed ellipses

$$Q = \pi \sum_{i=1}^{n} a_i b_i, \quad (16)$$

and determine the metric parameters of the coverage domain S_0—rectangle with sides $a_0 = \sqrt{2Q}$, $b_0 = \frac{\sqrt{2Q}}{2}$.

Thus, we form 50 tests that differ from each other in the number n of ellipses, their semi-axis (a_i, b_i), $i \in J_n$ and the sizes (a_0, b_0) of the coverage domain.

The next step is to generate ellipse placement parameters, which we also choose for each test evenly distributed on the intervals $(-\frac{a_0}{2}, \frac{a_0}{2})$, $(-\frac{b_0}{2}, \frac{b_0}{2})$, $(0, \pi)$, respectively.

Based on the generated initial data, we calculate the area $\mu(\Omega(g))$ of a complex object $\Omega(g)$, the structure of which is determined by Formula (3). The *Shapely* library was used to build $\Omega(g)$ and calculate its area.

Figure 1 shows the dependence of the average time of calculating the area $\mu(\Omega(g))$ on the number of ellipses S_i, $i \in J_n$ covering the rectangle S_0. Averaging was performed on 20 independent tests.

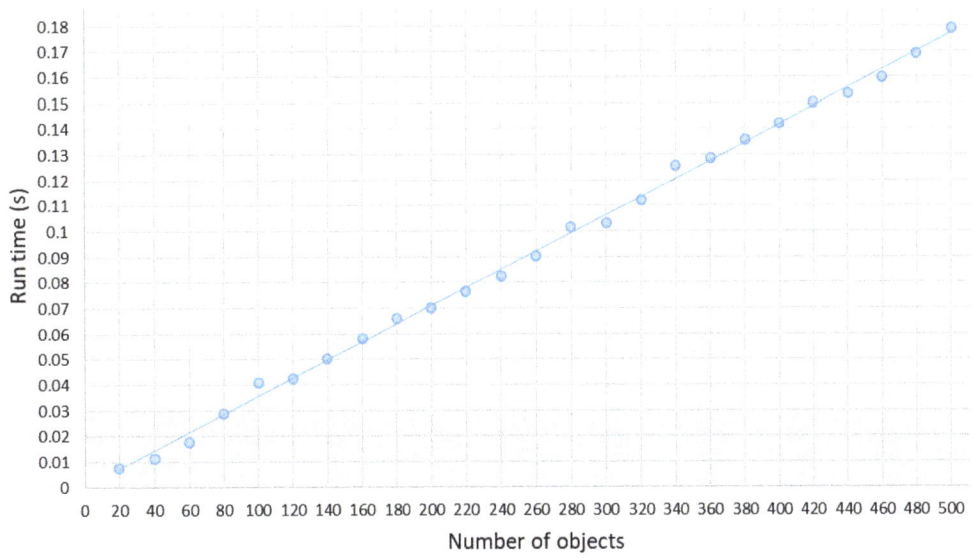

Figure 1. Dependence of the time of calculation of a complex object area on the number of objects.

The next experiment is to study the time of obtaining a local solution to problem (9). For local optimization, the *SciPy* package was chosen, which implements the *BFGS* method [37] using first order differences. The gradient was estimated by Formula (14). For each of the previously formed test tasks, the placement parameters $\mathbf{p}^i = (x_i, y_i, \theta_i)$, $i \in J_n$ randomly generated at the previous stage during the formation of a complex object were chosen as a starting point.

Figure 2 shows the dependence of the average time to obtain a local solution on the number of ellipses. Averaging was carried out over 10 independent tests. Formula (14) was used instead of (11) to estimate first-order differences when calculating the gradient of function $\omega_\Omega(\mathbf{g})$.

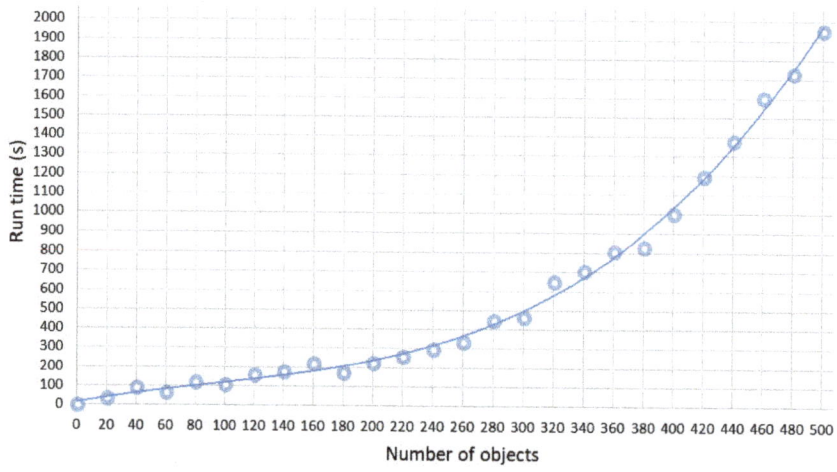

Figure 2. Dependence of local optimization run time on the number of objects.

To substantiate this approach, we studied the dependence of the formation time of the matrix of pairwise intersections (12) on its dimension. Figure 3 illustrates such dependence.

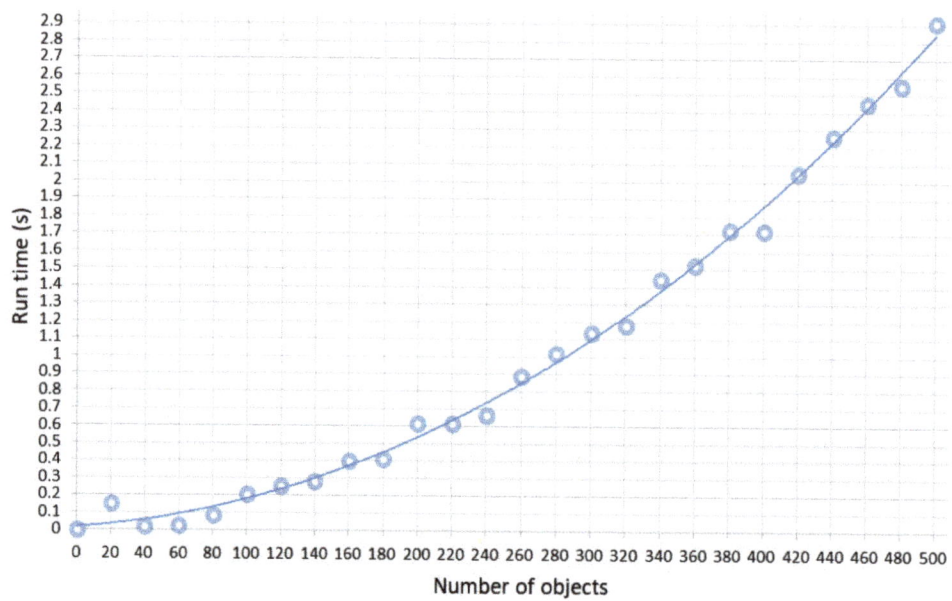

Figure 3. Dependence of the formation time of the matrix M on its dimension.

The data shown in Figure 3 allows us to estimate the average time for calculating the area of intersection of two objects. To do this, you need to sum all the elements of the matrix and divide by their number. Analyzing the obtained results, it can be argued that the estimation of the gradient using the matrix M requires much less time than by calculating the function $\mu(\Omega(g))$. Indeed, with n=100, the time for calculating the increment of a function by Formula (11) is $0.04 \times 2 = 0.08$ s, and the average time for the formation of the upper triangular matrix M is 0.2 s. Therefore, to calculate the area of intersection of each two objects, $0.2 \times 2/(100 \times 101) = 0.000039604$ s is required. Since to estimate $\Delta_k^i \omega_\Omega(\hat{g})$ using Formula (14) it is necessary to sum the areas of pairwise intersections of one object with the other 100, this will require 0.0039604 s. You can see the same when n = 500. The function $\omega_\Omega(\hat{g})$ calculation time is 0.18 s and the matrix M formation time is 2.7 s. The function increment $\Delta_k^i \omega_\Omega(\hat{g})$ according to Formula (14) will be calculated approximately in $2.7 \times 2/501 = 0.01078$ s.

Let us apply the described approach to solving the following test problems of maximum coverage. Let the covering objects be ellipses with semi-axis (a_i, b_i), $i \in J_n$, $n = 100$ listed in Table A1 (Appendix A).

The problems of unconditional local optimization (9) for n = 30, 50, 75, 100 was solved. Squares with sides 30, 40, 70, and 70 for n = 30, 50, 70, and 100, respectively, were chosen as the coverage domain. *BFGS* method using first-order differences was used. The gradients was estimated by Formula (14). The starting point for *BFGS* method was chosen by random generation of ellipse placement parameters by analogy with the experiments described earlier.

Figures 4–7 show the initial location of the ellipses in the coverage domain and the placement obtained as a result of local optimization for n = 30, 50, 75, and 100. The run times to solve the problems are 22 s, 51 s, 79 s, and 103 s, respectively.

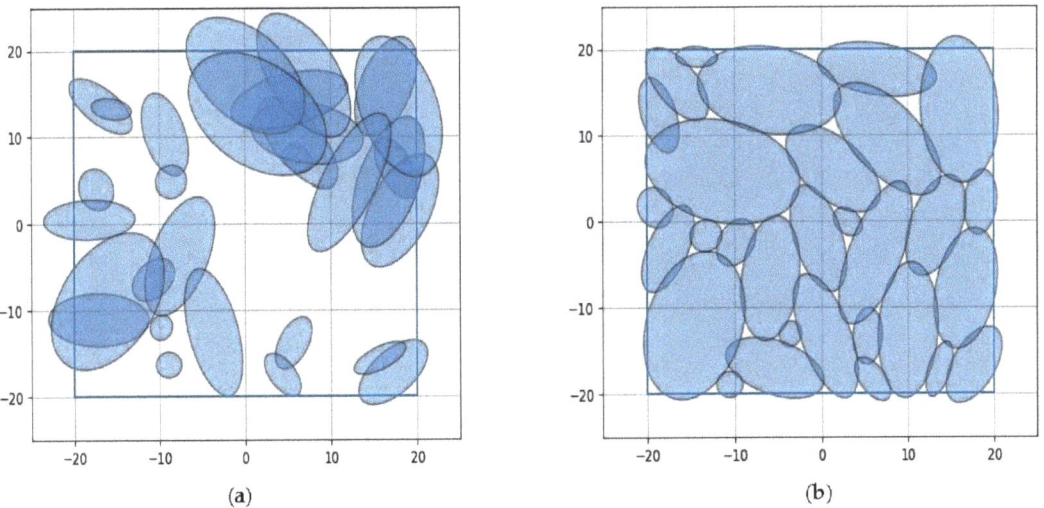

Figure 4. The continuous MCLP of 30 ellipses: (**a**) starting point and (**b**) local optimum.

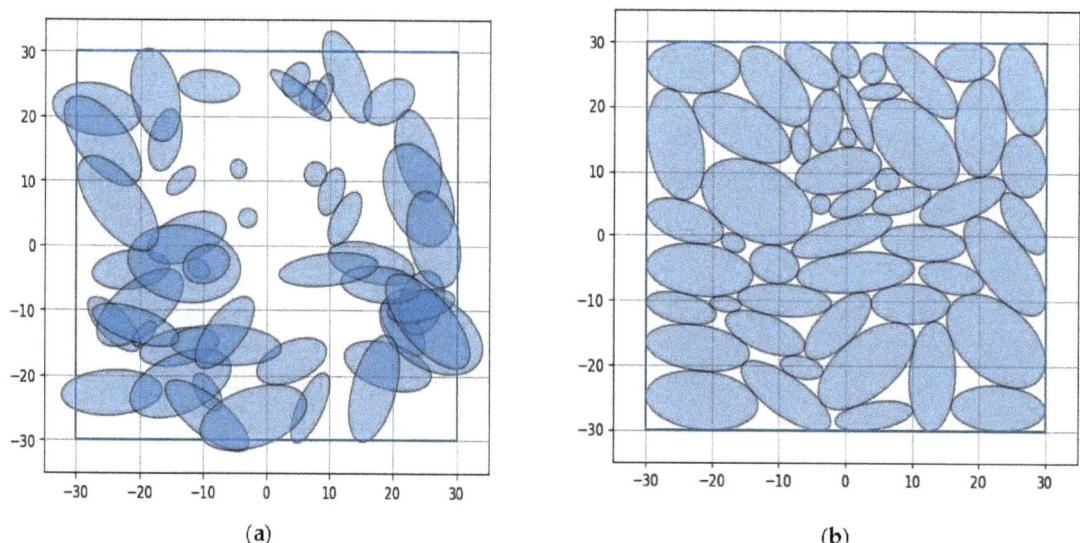

Figure 5. The continuous MCLP of 50 ellipses: (**a**) starting point and (**b**) local optimum.

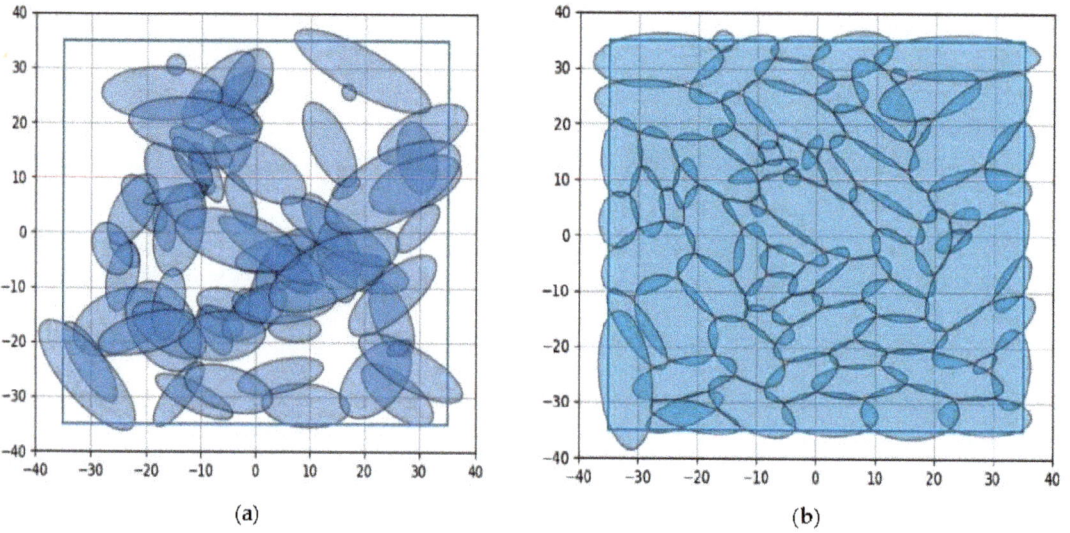

Figure 6. The continuous MCLP of 75 ellipses: (**a**) starting point and (**b**) local optimum.

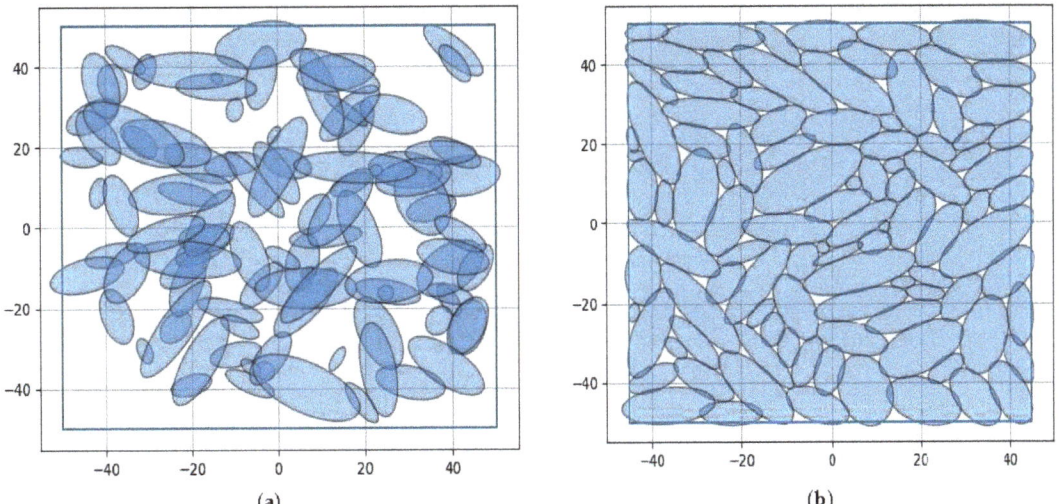

Figure 7. The continuous MCLP of 100 ellipses: (**a**) starting point and (**b**) local optimum.

Let us pay special attention to the test problems when n = 50 and n = 75. At n = 50, due to the large size of the coverage domain, such an placement of ellipses was obtained in which they do not intersect and are located inside the area. This is a global solution to problem (9), which corresponds to the so-called packing configuration [28,29]. For n = 75, each point of the coverage domain belongs to at least one of the covering objects. This is also a global solution to problem (9), and the resulting location defines the so-called covering configuration [28,29].

5. Discussion

The maximum coverage location problem is related to the determination of the placement of objects in order to cover as much as possible of the domain served by them. Such problems are formulated in discrete, continuous, and mixed statements, taking into account their geometric features.

The choice of one or another model depends on the restrictions on the location of objects and the properties of the service area. The classical discrete setting assumes that both object locations and service areas are isolated points. The weakening of the discreteness conditions leads to continuous and mixed formulations.

This article is perhaps one of the first to raise questions of formalization of completely continuous MCLP. In this case, both the locations and the service area are continual and suggest the possibility of their continuous transformation.

Of course, by discretizing the coverage area, for example, by grid methods, one can reduce continuous problems to discrete formulations. However, in this case, we will rather talk about the method of solution, and not about an adequate mathematical statement of the problem.

The geometric features of the continuous MCLP are related to both the shape of the covering objects and the service area. In turn, the problems go beyond the class of classical coverage problems, in which each point of the domain must belong to at least one of the covering objects. Therefore, continuous MCLP require the construction of special mathematical models and optimization methods that take into account their specifics.

The continuous MCLP model as a non-linear unconstrained optimization problem opens up prospects for using both modern effective methods for solving them and existing software in the form of powerful packages for various computer platforms.

The conducted studies allowed us to approach the analysis of the mathematical model of the problem from a unified standpoint. On the one hand, knowing the equations of general position of geometric objects $S_i(\mathbf{g}^i)$, $i \in \mathbf{J}_n^0$, used in the formation of a complex object $\Omega(\mathbf{g})$, one can write down the equation of its boundary. Therefore, formally we will find the analytical form of the function $\omega_\Omega(\mathbf{g}) = \mu(\Omega(\mathbf{g}))$ depending on the generalized variables $\mathbf{g}$. As a result, we will obtain an expression in the form of an integral depending on a vector parameter, for the calculation of which we will have to use numerical methods. This approach is very time consuming and is associated with the consideration of various options for the acceptable location of objects.

The approach proposed in the paper is devoid of these shortcomings. All the difficulties associated with the analytical specification of the function $\omega_\Omega(g)$ are overcome by using computer geometry packages that work great with geometric objects of complex spatial shape. As a result, it takes milliseconds to calculate the function value $\omega_\Omega(\hat{\mathbf{g}})$ for the fixed $\hat{\mathbf{g}}$, which is confirmed by the experiments described in the Section 4. This justifies the possibility of using metaheuristic methods of global optimization [38,39]. The efficiency of methods increases significantly if it is possible to use local optimization methods. The simplest is the multistart scheme for generation $\Omega(\hat{\mathbf{g}})$ with subsequent local optimization of the function $\omega_\Omega(\mathbf{g})$, choosing $\hat{\mathbf{g}}$ as the starting point. With limited time, the data illustrated in Figure 2 allows an estimate of the number of iterations in the multistart method. If temporary resources allow, hybrid methods can be proposed, where local optimization is applied after improvements have been received.

The geometrical features of the continuous MCLP model have made it possible to significantly increase the efficiency of local optimization methods using first order differences. This is confirmed by the data shown in Figure 3.

In general, the proposed approach opens up prospects for the development of new methods for solving both packing and covering problems. Indeed, in the course of the experiments, we obtained such locations of objects that corresponded to the packing configuration (Figure 5b) and covering configuration (Figure 6b). Thus, the introduction of additional conditions on the value of the function will allow us to propose new mathematical models for classical packing and covering problems and methods for their solution. In particular, the constraint

$$\omega_\Omega(\mathbf{g}) = \sum_{i=1}^{n} \mu(S_i), \qquad (17)$$

is the condition for both non-intersection of objects $S_i(\mathbf{g}^i)$, $i \in \mathbf{J}_n$, and their inclusion inside the container S_0.

The constraint

$$\omega_\Omega(\mathbf{g}) = \mu(S_0), \qquad (18)$$

is a condition for covering the domain S_0 by a family of geometric objects $S_i(\mathbf{g}^i)$, $i \in \mathbf{J}_n^0$.

Note that the choice of the *BFGS* method for local optimization was based on high user ratings. Of course, further studies of continuous MCLP require a comparative analysis of modern optimization methods, in particular [40]. Potential decision improvements can be obtained by using a novel optimization scheme, such as deep learned recurrent type-3 fuzzy system [41].

Another promising direction is the study of optimization problems of packing and coverage in domains with variable metric parameters. The results described in this article are directly related to this problem as well.

Author Contributions: Conceptualization, S.Y.; methodology, S.Y. and O.K.; software, O.K. and D.P.; validation, O.K. and D.P.; formal analysis, S.Y. and O.K.; investigation, O.K. and D.P.; resources, S.Y., O.K. and D.P.; writing—original draft preparation, S.Y.; writing—review and editing, S.Y. and O.K.; visualization, D.P.; supervision, S.Y.; project administration, S.Y. All authors have read and agreed to the published version of the manuscript.

Funding: The study was funded by the Ministry of Education and Science of Ukraine in the framework of the research project on the topic "Technologies, tools for mathematical modeling, optimization and systematic analysis of coverage problems in space monitoring systems".

Institutional Review Board Statement: Not applicable.

Informed Consent Statement: Not applicable.

Data Availability Statement: Generated data and test tasks are used.

Conflicts of Interest: The authors declare no conflict of interest.

Appendix A

Table A1. Metrical parameters of ellipses.

i	a_i	b_i	i	a_i	b_i	i	a_i	b_i	i	a_i	b_i
1	1.5	1.3	26	1.5	1.5	51	4.5	2.5	76	4.5	2.0
2	1.9	1.4	27	1.9	1.8	52	4.9	2.8	77	4.9	2.2
3	2.4	2.0	28	2.4	1.2	53	5.4	2.2	78	5.4	3.2
4	2.9	2.0	29	2.9	1.5	54	5.9	3.5	79	5.9	3.5
5	3.3	1,7	30	3.3	1.3	55	6.3	2.3	80	6.3	3.3
6	3.8	1.9	31	3.8	2.9	56	6.8	3.5	81	6.8	2.8
7	4.5	2.0	32	4.5	3.0	57	7.5	3.3	82	7.5	4.0
8	4.9	2.5	33	4.9	2.5	58	7.9	4.5	83	7.9	3.4
9	5.0	2.4	34	5.0	3.4	59	8.0	3.4	84	8.0	4.4
10	5.4	2.3	35	5.4	2.3	60	8.4	2.7	85	8.4	4.1
11	5.9	3.1	36	5.9	2.1	61	8.9	4.1	86	8.9	3.3
12	6.1	3.0	37	6.1	1.5	62	9.1	3.5	87	9.1	4.6
13	6.4	2.8	38	6.4	2.8	63	9.4	3.8	88	9.4	3.2
14	6.6	3.5	39	6.6	2.9	64	9.6	4.5	89	9.6	3.5
15	7.0	3.0	40	7.0	4.0	65	10.1	4.0	90	10.1	5.2
16	7.1	3.5	41	7.1	2.5	66	10.4	2.5	91	10.4	4.5
17	7.3	3.3	42	7.3	3.7	67	10.7	3.7	92	10.7	5.7
18	7.6	2.9	43	7.6	3.9	68	11.0	4.9	93	11.0	3.9
19	7.9	3.5	44	7.9	2.5	69	11.4	3.5	94	11.4	4.5
20	8.1	4.0	45	8.1	3.0	70	11.8	5.0	95	11.8	4.0
21	8.3	5.0	46	8.3	4.0	71	12.1	5.0	96	12.1	4.0
22	8.5	4.5	47	8,5	3.5	72	12.5	4.5	97	12.5	3.5
23	8.8	5.5	48	8.8	4.5	72	13.1	5.5	98	13.1	4.5
24	8.8	3.1	49	8.8	4.1	74	13.6	6.1	99	13.6	4.1
25	9.0	5.9	50	9.0	3.9	75	14.0	4.0	100	14.0	3.9

References

1. Owen, S.H.; Daskin, M.S. Strategic Facility Location: A Review. *Eur. J. Oper. Res.* **1998**, *111*, 423–447. [CrossRef]
2. Hamacher, H.W.; Drezner, Z. (Eds.) *Facility Location: Applications and Theory*; Springer: Berlin, Germany, 2002.
3. Church, R.L.; Murray, A.T. *Business Site Selection, Location Analysis, and GIS*; Wiley: New York, NY, USA, 2009.
4. Murray, A.T. Advances in location modeling: GIS linkages and contributions. *J. Geogr. Syst.* **2010**, *12*, 335–354. [CrossRef]
5. Eiselt, H.A.; Marianov, V. (Eds.) *Foundations of Location Analysis*; Springer: New York, NY, USA, 2011.
6. Murray, A.T. Maximal Coverage Location Problem: Impacts, Significance, and Evolution. *Int. Reg. Sci. Rev.* **2016**, *39*, 5–27. [CrossRef]

7. Church, R.L.; ReVelle, C. The maximal covering location problem. *Pap. Reg. Sci. Assoc.* **1974**, *32*, 101–118. [CrossRef]
8. Church, R.L. The planar maximal covering location problem. *J. Reg. Sci.* **1984**, *24*, 185–201. [CrossRef]
9. Hochbaum, D.S. Approximating Covering and Packing Problems: Set Cover, Vertex Cover, Independent Set, and Related Problems. In *Approximation Algorithms for NP-Hard Problems*; Hochbaum, D.S., Ed.; PWS Publishing Co.: Boston, MA, USA, 1996; pp. 94–143.
10. Bansal, M.; Kianfar, K. Planar Maximum Coverage Location Problem with Partial Coverage and Rectangular Demand and Service Zones. *INFORMS J. Comput.* **2017**, *29*, 152–169. [CrossRef]
11. Murray, A.T.; Matisziw, T.; Wei, H.; Tong, D. A Geocomputational Heuristic for Coverage Maximization in Service Facility Siting. *Trans. GIS* **2008**, *12*, 757–773. [CrossRef]
12. Wei, R.; Murray, A.T. Continuous space maximal coverage: Insights, advances and challenges. *Comput. Oper. Res.* **2015**, *62*, 325–336. [CrossRef]
13. Davari, S.; Zarandi, M.H.F.; Hemmati, A. Maximal covering location problem (MCLP) with fuzzy travel times. *Expert Syst. Appl.* **2011**, *38*, 14535–14541. [CrossRef]
14. Allahyari, M.Z.; Azab, A. Mathematical modeling and multi-start search simulated annealing for unequal-area facility layout problem. *Expert Syst. Appl.* **2018**, *91*, 46–62. [CrossRef]
15. Blanco, V.; Gázquez, R. Continuous maximal covering location problems with interconnected facilities. *Comput. Oper. Res.* **2021**, *132*, 105310. [CrossRef]
16. Murray, A.T.; Tong, D. Coverage optimization in continuous space facility siting. *Int. J. Geogr. Inf. Sci.* **2007**, *21*, 757–776. [CrossRef]
17. Murray, A.T.; O'Kelly, M.E.; Church, R.L. Regional service coverage modeling. *Comput. Oper. Res.* **2008**, *35*, 339–355. [CrossRef]
18. Tong, D.; Murray, A.T. Maximising coverage of spatial demand for service. *Pap. Reg. Sci.* **2009**, *88*, 85–97. [CrossRef]
19. Matisziw, T.C.; Murray, A.T. Siting a facility in continuous space to maximize coverage of a region. *Socio-Econ. Plan. Sci.* **2009**, *43*, 131–139. [CrossRef]
20. Coll, N.; Fort, M.; Sellarés, J.A. On the overlap area of a disk and a piecewise circular domain. *Comput. Oper. Res.* **2019**, *104*, 59–73. [CrossRef]
21. Mount, D.M.; Silverman, R.; Wu, A.Y. On the Area of Overlap of Translated Polygons. *Comput. Vis. Image Underst.* **1996**, *64*, 53–61. [CrossRef]
22. Coll, N.; Fort, M.; Saus, M. Coverage area maximization with parallel simulated annealing. *Expert Syst. Appl.* **2022**, *202*. [CrossRef]
23. Cheng, S.W.; Lam, C.K. Shape matching under rigid motion. *Comput. Geom. Theory Appl.* **2013**, *46*, 591–603. [CrossRef]
24. Stoyan, Y.G.; Patsuk, V. Covering a compact polygonal set by identical circles. *Comput. Optim. Appl.* **2010**, *46*, 75–92. [CrossRef]
25. A Gerrard, R.; Church, R.L.; Stoms, D.M.; Davis, F.W. Selecting conservation reserves using species-covering models: Adapting the ARC/INFO GIS. *Trans. GIS* **1997**, *2*, 45–60. [CrossRef]
26. García-Palomares, J.C.; Gutiérrez, J.; Latorre, M. Optimizing the location of stations in bike-sharing programs: A GIS approach. *Appl. Geogr.* **2012**, *35*, 235–246. [CrossRef]
27. Longley, P.A.; Goodchild, M.F.; Maguire, D.J.; Rhind, D.W. *Geographic Information System and Science*, 4th ed.; John Wiley: Chichester, UK, 2015.
28. Stoyan, Y.G.; Yakovlev, S.V. Configuration Space of Geometric Objects. *Cybern. Syst. Anal.* **2018**, *54*, 716–726. [CrossRef]
29. Yakovlev, S.V. On Some Classes of Spatial Configurations of Geometric Objects and their Formalization. *J. Autom. Inf. Sci.* **2018**, *50*, 38–50. [CrossRef]
30. Yakovlev, S.V. Formalizing Spatial Configuration Optimization Problems with the Use of a Special Function Class. *Cybern. Syst. Anal.* **2019**, *55*, 581–589. [CrossRef]
31. Rvachev, V.L. *Theory of the R-Function and Some of Its Applications*; Naukova Dumka: Kyiv, Ukraine, 1982; 552p.
32. Stoyan, Y.; Scheithauer, G.; Gil, N.; Romanova, T. Φ-functions for complex 2D-objects. *Q. J. Oper. Res.* **2004**, *2*, 69–84.
33. Bennell, J.; Scheithauer, G.; Stoyan, Y.; Romanova, T. Tools of mathematical modeling of arbitrary object packing problems. *Ann. Oper. Res.* **2010**, *179*, 343–368. [CrossRef]
34. Stoyan, Y.G.; Romanova, T.; Scheithauer, G.; Krivulya, A. Covering a polygonal region by rectangles. *Comput. Optim. Appl.* **2011**, *48*, 675–695. [CrossRef]
35. Pankratov, A.; Romanova, T.; Litvinchev, I.; Marmolejo-Saucedo, J.A. An Optimized Covering Spheroids by Spheres. *Appl. Sci.* **2020**, *10*, 1846. [CrossRef]
36. Gillies, S. The Shapely User Manual. Available online: https://shapely.readthedocs.io/en/stable/manual.html (accessed on 29 April 2022).
37. Fletcher, R. *Practical Methods of Optimization*, 2nd ed.; John Wiley Sons: New York, NY, USA, 2013; ISBN 978-0-471-91547-8.
38. Glover, F.; Sorensen, K. Metaheuristics. *Scholarpedia* **2015**, *10*, 6532. [CrossRef]
39. Salhi, S.; Thompson, J. An Overview of Heuristics and Metaheuristics. In *The Palgrave Handbook of Operations Research*; Salhi, S., Boylan, J., Eds.; Palgrave Macmillan: Cham, Switzerland, 2022. [CrossRef]
40. Feng, W.; Salgado, A.J.; Wang, C.; Wise, S.M. Preconditioned steepest descent methods for some nonlinear elliptic equations involving p-Laplacian terms. *J. Comput. Phys.* **2017**, *334*, 45–67. [CrossRef]
41. Cao, Y.; Raise, A.; Mohammadzadeh, A.; Rathinasamy, S.; Band, S.S.; Mosavi, A. Deep learned recurrent type-3 fuzzy system: Application for renewable energy modeling/prediction. *Energy Rep.* **2021**, *7*, 8115–8127. [CrossRef]

Article

Concept of High-Tech Enterprise Development Management in the Context of Digital Transformation

Yurii Pronchakov [1], Oleksandr Prokhorov [2,*] and Oleg Fedorovich [2]

1. Software Engineering and Business Faculty, National Aerospace University "Kharkiv Aviation Institute", 61070 Kharkiv, Ukraine; pronchakov@gmail.com
2. Computer Sciences and Information Technologies Department, National Aerospace University "Kharkiv Aviation Institute", 61070 Kharkiv, Ukraine; oe.fedorovich@gmail.com
* Correspondence: o.prokhorov@khai.edu

Abstract: The purpose of this article is to check and identify management gaps that lead to the formation of digitalization problems in enterprises in the context of Industry 4.0 and to offer a conceptual approach to managing the development of high-tech enterprises in digital transformation. The paper substantiates the concept of digital transformation management in a high-tech enterprise based on interdependent adaptive systems for planning digital transformation processes, monitoring, and change management. The paper considers the idea of the Industry 4.0 concept and presents principal technologies and tools that contribute to the gradual transition to digital transformation. It is determined that digital transformation is a process of transition to digital business, which involves the use of digital technologies to change business processes in the company and provision of new opportunities for additional income and development prospects. A conceptual model of enterprise competitiveness formation in the process of digital transformation has been developed, which includes organizational and economical digital tools for sustainable development of high-tech enterprises and synergies from the organization of new forms of digital interaction. The proposed methodology for managing the development of high-tech enterprises in the context of digital transformation is based on formation of an ecosystem model of decentralization in a single distributed digital space, based on interconnected adaptive systems of planning, monitoring, and change management, and, on the basis of modeling and forecasting of complex manufacturing and logistics processes of high-tech industries, it allows effective implementation of the innovative order portfolio in the short term and with limited opportunities while coordinating the priorities of the business strategy and strategy of digital transformation of high-tech enterprises.

Keywords: digital transformation; high-tech enterprises; Industry 4.0

Citation: Pronchakov, Y.; Prokhorov, O.; Fedorovich, O. Concept of High-Tech Enterprise Development Management in the Context of Digital Transformation. *Computation* **2022**, *10*, 118. https://doi.org/10.3390/computation10070118

Academic Editor: Demos T. Tsahalis

Received: 30 May 2022
Accepted: 8 July 2022
Published: 10 July 2022

Publisher's Note: MDPI stays neutral with regard to jurisdictional claims in published maps and institutional affiliations.

Copyright: © 2022 by the authors. Licensee MDPI, Basel, Switzerland. This article is an open access article distributed under the terms and conditions of the Creative Commons Attribution (CC BY) license (https://creativecommons.org/licenses/by/4.0/).

1. Introduction

The growth of global competition has led to significant changes in the operations of manufacturing companies. Many past concepts have been revised since the fourth industrial revolution: business models, collaboration, user interfaces, value chains, and even the traditional automation pyramid, which are currently undergoing major changes [1]. In order to remain competitive, manufacturing companies must constantly increase the efficiency and performance of their production processes.

Increasing digitalization, awareness, global initiatives, and regulations are putting pressure on the high-tech industry to shift towards sustainable development. Digital transformation is one of the most important components of future product development [2]. New opportunities will result in the minimization of the human factor in many manufacturing and logistics processes, a fundamentally new quality of management decisions at key levels of government through extensive use of data and digital models, and large-scale integration of end-to-end manufacturing and logistics chains [3].

Digital transformation of business processes and business models is a necessary and relevant stage in the development of the high-tech industry; in fact, it is the transformation of high-tech industry into digital industry based on the development and application of digital platforms; development and application of digital duplicates; implementation of the transition to cyber-physical systems [4]. Moreover, the digital form of interaction of businesses poses new tasks that are of a manufacturing, technological, organizational, and managerial nature, which require the fastest and most effective solution [5]. Paper [6] notes that the main factors of development are human capital, access to information space, and the ability to develop and implement innovations.

Digital transformation is the introduction of modern digital technologies into the business processes of an enterprise at all levels. This approach implies not only the installation of modern hardware or software but also fundamental changes in management approaches. Digital transformation provides the fullest disclosure of the potential of digital technologies through their use in all aspects of business processes, products and services, and approaches to decision making. Digital transformation can be considered only at the intersection of all three dimensions (formulated business task, data availability, and technology itself).

In our study, it is important to define digital transformation as a process of integrating digital technologies into all aspects of business activities, which requires fundamental changes in technology, culture, operations, and principles of creating new products and services.

The implementation of innovation processes, especially large-scale ones as in the concept of digital transformation, is very challenging and entails high costs. Identifying and understanding the challenges that may arise in the process of digital transformation is crucial for successful implementation. Further, the paper considers the existing approaches to analyzing and managing digital transformation.

One of the approaches is Digitization Piano, created by the Global Center for Digital Business Transformation (Lausanne, Switzerland) [7]. Similarly to seven notes, it distinguishes seven categories of transformation: business model ("how you make money"), structure ("how you are organized"), people ("the people who work for you"), processes ("how you do things"), IT capability ("how you manage information"), offerings ("your products and services"), and engagement ("how you engage stakeholders"). For this concept, it is important to define the gap between the current state and needs. The Digitization Piano technique, which divides the organizational value chain into seven critical elements, shows that the chances of successful transformation increase if the organization addresses one element at a time, i.e., plays chords rather than keys.

German economists D. Schallmo and C. Williams developed a sequence of stages (phases) of digital transformation of business models, which includes the following stages [8]: Digital Reality, which defines the existing business model of the company together with a review of customer requirements; Digital Ambition, which defines the main goals of the transformation; Digital Potential, which identifies best practices and factors that contribute to the development of digital transformation; Digital Fit, where the analysis and evaluation of options for a new digital business model occur; Digital Implementation, which includes refinement and implementation of the developed digital business model.

The degree of implementation of digitalization elements in an enterprise is one of the key criteria for increasing the level of its technological maturity. Multiple studies are devoted to creating models and evaluating the digital maturity of enterprises [9–11]. Paper [11] proposes a model for assessing the company's maturity within Industry 4.0 in 8 dimensions and 65 parameters. Indeed, "digital" behavior depends on the company's competencies and the level of maturity of the enterprise's digital management system, which is dynamic and needs to be improved in order to move to a higher level.

It should be noted that digital awareness, i.e., the company's knowledge of its digital assets and potential, allows the most accurate selection of priority areas of digital transformation with the greatest efficiency and profitability. Available digital assets provide an opportunity to obtain the predicted key behavior of the enterprise in carrying out its activities in the context of digital transformation, which is of special importance. Paper [12] presents the development of a set of factors for assessing the commercial potential of different technological industries based on multi-criteria decision-making methods.

Thus, if we systematize and generalize the main directions of research on the digital transformation of enterprises, it is possible to identify the following: strategies for digital transformation [13]; sustainability and circular economy [2,14]; models of digital maturity of enterprises [15,16]; digital capability models [17,18]; strategic roadmaps [19]; formation of new business models [20]; reforming organizational structures [4].

The analysis demonstrated that most researchers understand the need for an integrated approach to managing enterprise development in a digital transformation, but some prefer either strategy, organizational structure, technological infrastructure, resources, culture, or workflows, and they do not provide ways to form a single system of principles, goals, objectives, and performance indicators for the management system of digitalization of the enterprise in the context of Industry 4.0.

The industrial concept of Industrial 4.0 is a large-scale multi-level organizational and technical system [21,22], which involves a transition to fully automated digital manufacturing controlled by real-time intelligent systems, where cyber-physical systems are integrated into a single information space using industrial Internet of Things technology. One of the forms of implementation of the concept of Industry 4.0 in paper [23] considers an intelligent manufacturing system, which is formed by adopting new models, new forms and new methodologies to transform the traditional manufacturing system into an intelligent one, whose structure includes the following categories: smart design, smart machines, smart monitoring, smart management, and smart planning. The Industrial 4.0 concept is based on the availability of complete and reliable digital data from smart devices, as well as new capabilities for their remote monitoring and processing, real-time manufacturing and management situations, and advanced analytics, wherein advanced methods of big data analytics using methods and artificial intelligence tools [24], machine learning, data mining and predictive analytics, etc. are the hallmarks of Industrial 4.0.

Many studies are related specifically to the technological aspects of implementing Industry 4.0. The following main technological trends in the field of digital transformation of industry can be identified: development and application of digital platforms; development and application of digital duplicates; implementation of the transition to cyber-physical systems.

However, the digital transformation or Industry 4.0 should not be limited to technology. It should also be linked to strategic changes in enterprises and organizational management, including strategy, structures, processes, resources, and culture [4,17,25].

The analysis showed that most of the research in this area is of an overview nature, describing various Industrial 4.0 technologies, how they are implemented in different enterprises, and what advantages they provide. It is quite clear that digitalization and the introduction of Industrial 4.0 technology is the foundation for building effective competitive management at modern high-tech enterprises. Unfortunately, most of these studies answer the question "what to do?" without providing the answer to "how to do it?".

In the context of striving to improve the economic stability of enterprises and increase product competitiveness, there is a gap between the need to implement digital transformation programs at enterprises and the imperfection of scientifically sound means of achieving this goal in the form of models and methods of digital transformation project portfolios management, which do not fully take into account the many interrelated resource flows, requirements, objectives, and behavior strategies of individual units of enterprises, as well as the dynamics of manufacturing and logistics processes.

Thus, this study solves the scientific and applied problem, which is the creation and implementation of methodological bases for the portfolio management of digital transformation projects in enterprises, taking into account strategic objectives, limited resources, and risks.

The purpose of this article is to check and identify management gaps that lead to the formation of digitalization problems in enterprises in the context of Industry 4.0 and to offer a conceptual approach to managing the development of high-tech enterprises in the digital transformation.

The scientific hypothesis of the study is based on the assumption that building an effective management system for projects and programs of digital development in enterprises, which will be based on predictive analytics and consist of interdependent adaptive systems of planning, monitoring, and change management, will allow more effective generation and implementation of a portfolio of digital transformation projects in enterprises and help increase the efficiency and competitiveness of enterprises. This paper reviews whether it is possible to indicate and identify the prerequisites and gaps that further result in specific problems of project implementation in the process of digital transformation of the enterprise.

In contrast to many review studies, the main advantage of the proposed approach is that, based on our practical experience in implementing development and digital transformation programs at high-tech enterprises, we will present both conceptual foundations and specific models: the model of forming the competitiveness of the enterprise in the process of digital transformation, which enables assessment of priorities of the digital policy of the enterprise; a multi-level model of managing digital transformation at the high-tech enterprise.

The following scientific methods are used in our research: system approach—during the formation of theoretical foundations of the methodology of project management and digital transformation programs at high-tech enterprises; methods of strategic planning—during the formation of the strategy of digital transformation management at the enterprise and its coordination with the business strategy of the enterprise; methods of expert evaluation and decision-making—during the assessment of the importance of digital policy priorities of the enterprise and selection of projects to the portfolio of development projects; methods of simulation modeling—during construction of the analysis model of energy distribution processes in the implementation of digital transformation projects.

2. The Concept of Managing the Development of High-Tech Enterprises in the Context of Digital Transformation

The formulation of management principles for the high-tech enterprise activity in the context of digital transformation on the basis of technologies of Industry 4.0 is an important initial stage of the creation of methodology and management system under new conditions of development. Our study of methodological aspects of management of the studied processes allowed us to determine the following basic principles of management and development of high-tech enterprises in the context of digital transformation in the following areas.

1 General Principles:

1.1 The principle of a systemic approach assumes that the planning and implementation of digital transformation processes shall be systemic. This principle implies that the processes of digital transformation shall be integrated into enterprise processes and coordinated with the financial, investment, organizational, production, and motivational policies of a high-tech enterprise. This approach allows taking into account all the necessary connections and interactions in the management and development of a high-tech enterprise, comprehensive assessment of the factors, and direct management mechanisms to achieve the goals of digital transformation of the enterprise. According to this principle, both the system as a whole and each of its subsystems shall perform the function of planning towards a single goal vector. Plans shall be linked through vertical integration and differentiation and through horizontal coordination. The more elements and levels the system has, the more profitable it is to plan them simultaneously and in interconnection.

1.2 The principle of continuity defines the process of digital transformation as a continuous process within a set cycle when the developed plans and projects replace each other. The principle applies primarily to plans of different time periods but also includes the relationship of planning with the forecasting, cycle, and sequence of stages of digital transformation.

1.3 The principle of adaptive management means that plans and the process of digital transformation can change their direction due to unforeseen circumstances. In general, this implies interconnected adaptive systems of planning, monitoring, and managing changes in the process of digital transformation, which are aimed at ensuring the long-term sustainable development of a high-tech enterprise through the efficient use of all digital assets and the potential of digital technologies, transfer of activities, and a new way of development.

1.4 The principle of innovation and advancement determines that the changes introduced during the digital transformation shall take into account the latest, newest innovative achievements of world scientific and technological progress, the world's leading technologies, which is a key feature for high-tech industries. In addition, the changes implemented at the high-tech enterprise shall provide a technical and technological breakthrough and competitive advantages for both the enterprise and the industry by being new not only for the enterprise but also for national and global markets.

1.5 The principle of openness and standardization stipulates that the openness of data for all participants in the digital transformation process is the basis for all participants to develop new solutions and services. The development of a high-tech enterprise on the basis of Industry 4.0 technology requires the development of a production system, production and logistics processes based on the relevant standards, which will ensure both implementations of digital transformation projects and compliance with relevant international standards of production and quality processes.

2 Organizational and Economic Principles:

2.1 The principle of a mature digital environment. Digital transformation is essentially a radical change in business strategy and corporate governance processes under the influence of digitalization. Therefore, "digital behavior" depends on the company's competencies and the level of maturity of the enterprise's digital management system, which is dynamic and needs to be improved in order to move to a higher level. Thus, such aspects of the enterprise as digital assets, management quality, organizational structure, competence, and staff motivation in the context of digitalization, availability of appropriate regulatory framework, and knowledge base in modern digital technologies have a strong impact on the results of digital transformation projects.

2.2 The principle of virtualization is due to the fact that requirements for the organization of new productions, taking into account the concept of Industry 4.0, have led to the creation of enterprises in the form of virtual productions, which are focused on creating new innovative products.

2.3 The principle of priority and purposefulness in making decisions on the implementation of certain projects of the digital transformation portfolio. Priorities are the basic concept and structural element of the strategy of digital transformation, and they represent the best directions and forms of enterprise management at each stage of its implementation. Investment resources are always limited, so this principle will ensure that a fixed amount of resources in the implementation of projects of high-tech enterprises will be spent in accordance with the achievement of strategic objectives and solving the most important problems of digitalization of high-tech enterprises. In addition, without a clear goal and results that can be expected to achieve it, any action in the implementation of the portfolio of digital transformation projects of the enterprise is bound to fail.

2.4 The principle of the ecosystem management model allows each participant in the digital ecosystem to optimize their business processes, diversify production, participate in the value chain of other participants, reduce transaction costs, and implement innovative projects, while receiving a powerful impact from interaction with other ecosystem participants.

2.5 The principle of synergy effect is the simultaneous achievement of the best technological, economic, organizational, environmental, and other end results from the merger of physical and digital technologies. The synergy effect of the portfolio of digital transformation projects means a situation when the resulting benefit from the implementation of the project portfolio exceeds the benefit from the implementation of portfolio projects separately. Industry 4.0 provides for end-to-end digitalization of all physical assets and their integration into the digital ecosystem together with partners involved in the value chain. Therefore, the paper considers the synergy effect of the organization of new forms of digital interaction in the form of ecosystems, such as the interaction of civil, defense, and space industries at the level of programs, technologies, and innovations.

3 Production and Technological Principles:

3.1 The principle of platforming determines that the digital platform is the technological basis for providing a set of new, specific services related to digital processes. This enables a significant reduction of the production cost of goods and services while accelerating the servicing of all types of economic activities of the enterprise.

3.2 The principle of design. In modern conditions of the industry, digital transformation can be observed as the formation of a new organization of manufacturing where the center of gravity is shifted towards design. Industrial development processes require the detailed planning and determination of necessary material, human, financial, and investment resources and the calculation of expected financial results and risks, respectively; they shall be implemented through innovation and investment projects and design mechanisms. On the other hand, it is the digital design and modeling of technological processes, objects, and products throughout the entire life cycle from idea to operation.

3.3 The principle of intellectualization determines the leading role of technologies of artificial intelligence, computer vision, and machine learning in the effective solution of digital transformation.

3.4 The principle of big data analytics. The application of various types of analytics, in particular predictive analytics for planning in a high-tech enterprise, obtaining, at the output, qualitatively new information used for process management.

3.5 The principle of integration. Processes of management and development of high-tech enterprises in the context of digital transformation shall be focused on the use of those processes, technologies, materials, and digital programs of protocols which will provide their maximum compatibility of technical and technological integration that would further provide the chance to solve integration problems of logistics networks, clusters, and integration with foreign economic markets at the level of other enterprises.

The structure of the proposed methodological foundations of management and development of high-tech enterprises in terms of digital transformation is shown in Figure 1. It is represented by principles, goals, objectives, models and methods, and possible directions of application.

The article is not devoted to the detailed calculations of partial mathematical models. Some of them can be found in [26–28]. We will continue to talk about the conceptual provisions of the approach.

The concept of managing the development of a high-tech enterprise in the context of digital transformation is shown in Figure 2.

Initially, the enterprise has a certain digital potential, is part of a certain digital ecosystem, and decides to implement a portfolio of digital transformation projects. The digital transformation strategy is based on digital awareness of the challenges and opportunities associated therewith, as well as strategic correspondences of production, technological, and business process chains. Strategic correspondences can be found in any part of the chain of processes: in R&D and technology, in the supply chain, supplier relations, manufacturing, sales, and marketing within administrative functions. For example, digital transformation projects, in which companies share technology and technological know-how, create valuable competitive advantages by reducing R&D costs, accelerating the launch of new high-tech products, and using new technological advances to implement new projects and programs.

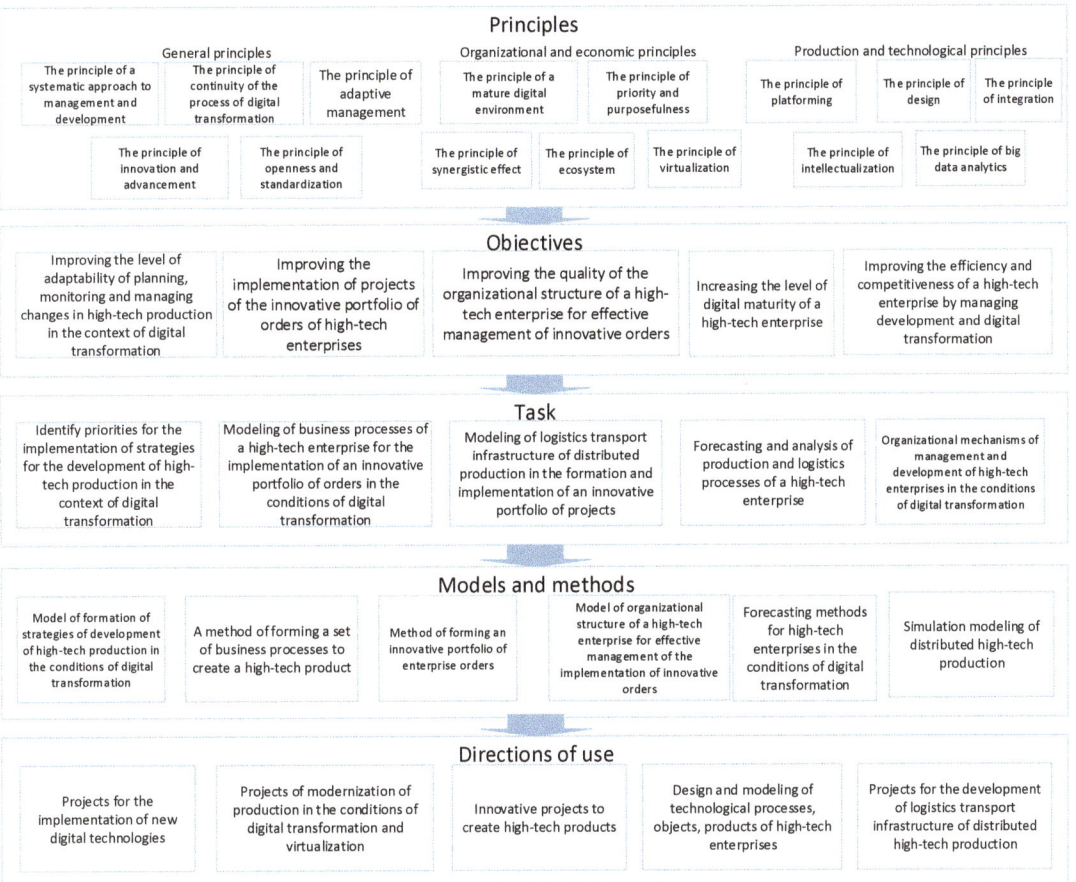

Figure 1. The structure of methodology of high-tech enterprise development management in the context of digital transformation. Source: own study.

Figure 2 show exactly how to manage the development of the enterprise during the implementation of the strategy of digital transformation.

It should be noted that digital awareness, i.e., the company's knowledge of its digital assets, allows the most accurate selection of priority areas of digital transformation with the greatest efficiency and profitability. Available digital assets provide an opportunity to obtain the predicted key behavior of the enterprise in carrying out its activities in the context of digital transformation, which is of special importance. As can be seen from the figure, in this case, the mechanisms of management, development, and transfer of competencies are involved, which allow the shift from the operation of the enterprise (business processes) in its basic version to the version taking into account the introduction of digital technologies and digital transformation projects.

In this case, various mechanisms implemented in the process of managing these digital competencies (e.g., implementation of digital platforms, transition to big data analytics, training of insufficient digital competencies of personnel, transfer of employees, technology, reputation, etc.) can achieve positive synergies.

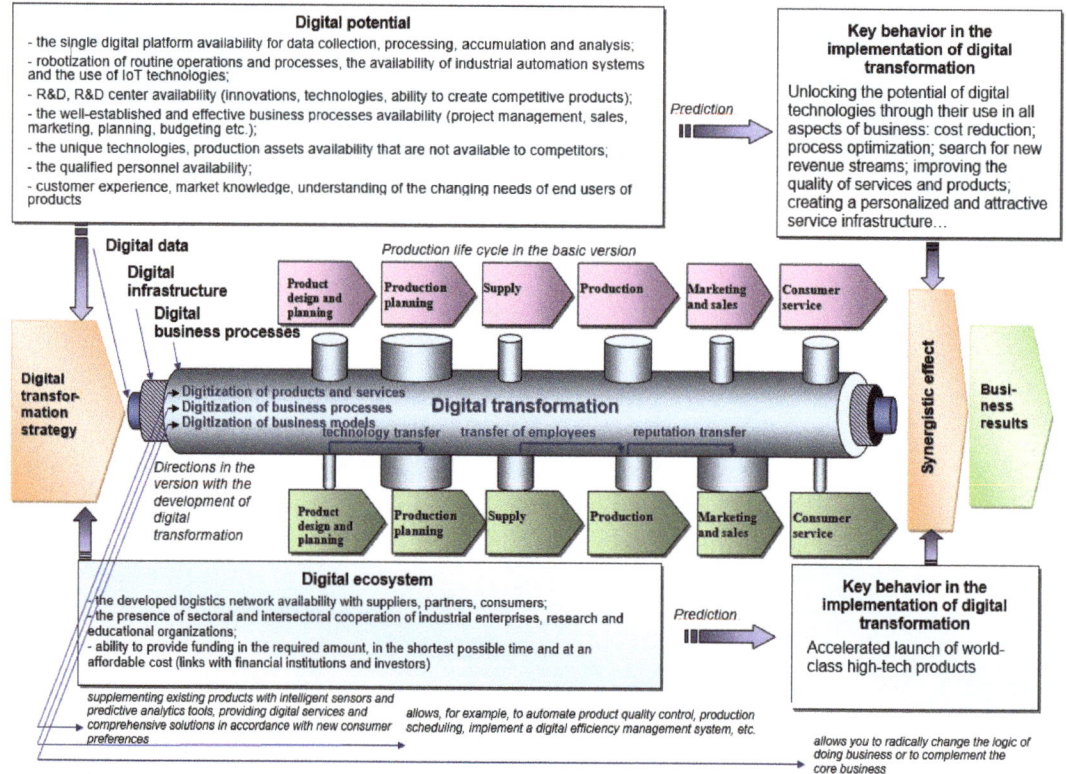

Figure 2. Concept of high-tech enterprise development management in the context of digital transformation. Source: own study.

The organizational and economic tools used in this process are designed to optimize business processes and create new business models for stakeholders in the ecosystem of high-tech enterprises in a single information environment.

As can be seen from Figure 2, the concept of the digital ecosystem, discussed in the first section, plays an important role.

It is the digital ecosystem that provides the necessary organizational and economic model of mutually beneficial operation of different stakeholders, taking into account the digital platform, which gives them opportunities to accelerate growth and reduce costs through synergies from multilateral cooperation based on common rules and principles.

The ecosystem mechanism of strategic development of high-tech industries shall provide sectoral and intersectoral cooperation of industrial enterprises, research, and educational organizations and provide stakeholders with tools and services for innovative projects and accelerated market launch of high-tech products.

The ecosystem management model allows each ecosystem stakeholder to optimize their business processes, diversify production, participate in the value chain of other stakeholders, reduce transaction costs, and implement innovative projects while receiving a powerful synergistic effect from interaction with other ecosystems stakeholders.

The paper determines the place of digital transformation management in the integrated project management system and development programs of a high-tech enterprise (Figure 3).

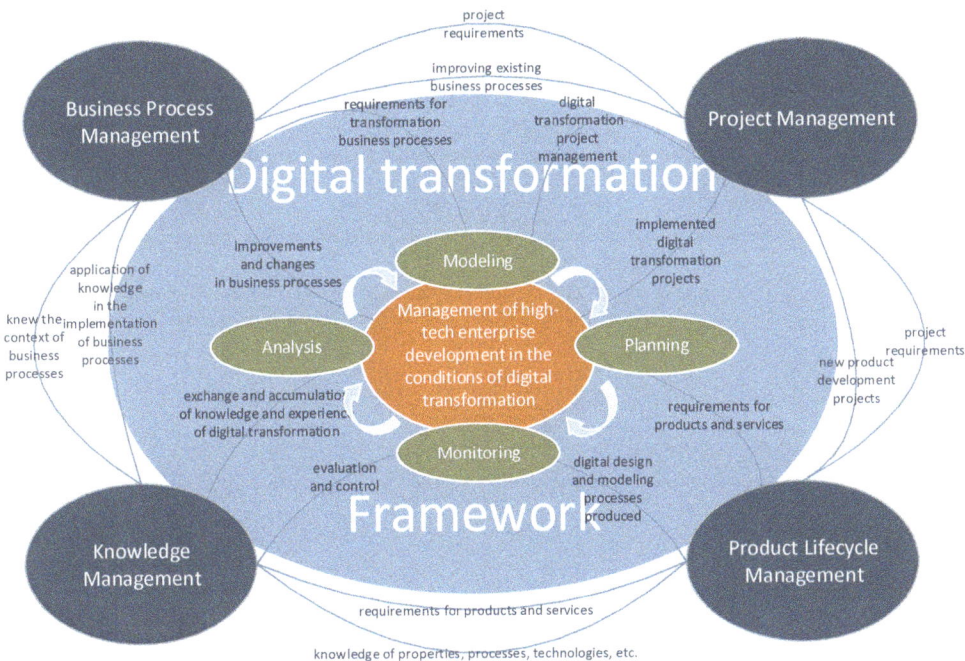

Figure 3. The digital transformation framework in the enterprise. Source: own study.

One of the main roles is assigned to the adaptation of the competency model to the demands and challenges of digital transformation due to the continuous improvement of existing business processes (Business Process Management, BPM) and the development and implementation of new business projects based on project management (PM) tools. Therein, requirements which are a basis for management of digital transformation are formulated on the basis of the current condition of the business processes of the enterprise. If the term "ability" is used to denote business processes, it turns out that the company will thus have many abilities (for example, IT readiness—the ability of the company to achieve its goals and objectives through the most efficient use/implementation of information technology). At the same time, the analysis of business processes that are strategically important, and their added value, enable the move to related key corporate competencies. For BPM, knowledge of a business process is a necessary (minimum) and sufficient set of information on the basis of which trained professionals can perform this business process. The knowledge of how to deploy, organize, and manage business processes most effectively in the new environment of digital transformation constitutes the competence of the enterprise.

Knowledge Management (KM) is the process of creating conditions for identification, storage and effective use of knowledge and information in the enterprise and its environment; part of whom are competencies of the enterprise and employees. Employee competencies are described as a set of requirements for their knowledge, skills, and qualities for a specific function, position, or role in a digital transformation project. The knowledge management strategy, in turn, aims to provide the necessary knowledge in a timely manner to those specialists of the enterprise who need this knowledge to perform new tasks related to the digitalization of processes and implementation of digital transformation projects. Product Lifecycle Management (PLM) involves bridging the gap between the processes of operation, manufacturing, and development through digital images of objects. One of the features of high-tech enterprises is the increased requirements for a set of digital competencies of employees, which is primarily due to the complexity of products created through the implementation of projects.

What about the mechanism for implementing the digital transformation strategy?

Within the framework of the methodology, it includes the following main components (Figure 4): formation and adoption of targeted programs to increase the level of digital maturity and efficiency of high-tech enterprises; formation of a modern system of digital standards and technologies; organization of work on the implementation of digital transformation projects and expansion of the practices of using modern digital technologies in the modernization and reconstruction of fixed assets; training and motivation of personnel in terms of digitalization.

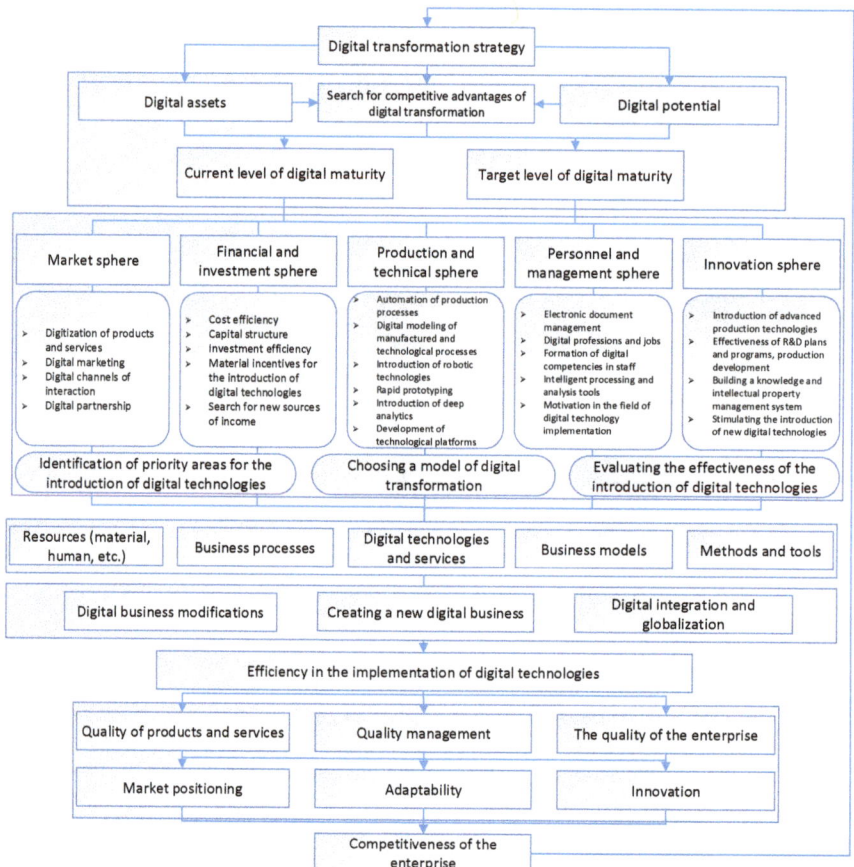

Figure 4. Conceptual model of enterprise competitiveness formation in the process of digital transformation. Source: own study.

The digital transformation strategy drives changes in business models and uses technology to create the opportunities needed by an enterprise to become a digital business and achieve certain performance criteria. Strategy determination is a key component of the transformation process, which ensures that the technology is implemented to support business goals.

The functional components of the digital strategy of the enterprise are the factors that determine the performance and efficiency of the strategy. In other words, digital transformation is not a short-term concept of expansion, so the lack of a strategy for digital transformation is one of the most serious obstacles to achieving the digital maturity of the enterprise [2].

It is proposed to distinguish several main types of digital transformation strategies:
- passive: opportunities for digital transformation are determined based on the actual level of development of digitalization of enterprise departments and their information needs;
- adaptive: opportunities are determined based on the expected information needs of a significant number of departments in accordance with the strategy of their development and financial capabilities of the enterprise;
- active: information needs of a significant part of departments are largely formed in accordance with ideas about the prospects of digital transformation;
- aggressive: information needs of the majority of departments are formed on the basis of their interests;
- breakthrough: based on innovative digital technology and platforms that are characterized by a high-risk level but involve the use of non-standard solutions to address the challenges facing the management of high-tech enterprises.

Digital transformation of the enterprise covers several areas, including the following: introduction of modern technologies and equipment or software in business processes; formation of the offer of new digital or digitized products and services; fundamental changes in approaches to enterprise and personnel management; transformation of interaction means within the enterprise, relations between employees; establishing external communications through digital communication channels, etc.

Digital transformation is a change of form of operations, restructuring of organizational structure, application of new business models, new sources and forms of income, attracting a wider range of consumers, bringing customer service to a new level, mixing areas of operation in new formats, including in the form of digital platforms [29]. Thus, the introduction of modern information technology is not actually a digital transformation but a means to achieve it. The three variable blocks in the transformation of business models are digital business modifications, the creation of new digital businesses, and digital integration and globalization.

The concept of digital transformation management in a high-tech enterprise is based on several interdependent adaptive systems: planning and formation, monitoring, and management of changes (Figure 5).

First, we need to conduct systematic modeling of strategic objectives of high-tech enterprise development, which forms a global goal of reform in the digital transformation and evaluates its implementation in the short term, taking into account the limited capabilities of the enterprise. This allows us to further determine the indicators and factors to achieve the nearest selected goal and justify and choose measures to improve the implementation of the new order portfolio to ensure the production of competitive products for the reforming enterprise.

Subsequently, it is necessary to simulate the business processes of a high-tech enterprise in order to fulfill an innovative order portfolio in the context of digital transformation. This takes into account the specifics of high-tech enterprises where complex equipment is created that requires a multi-component presentation of its architecture.

The next step is to model the organizational structure of the enterprise, which is sufficient and necessary for the effective implementation of the innovative order portfolio.

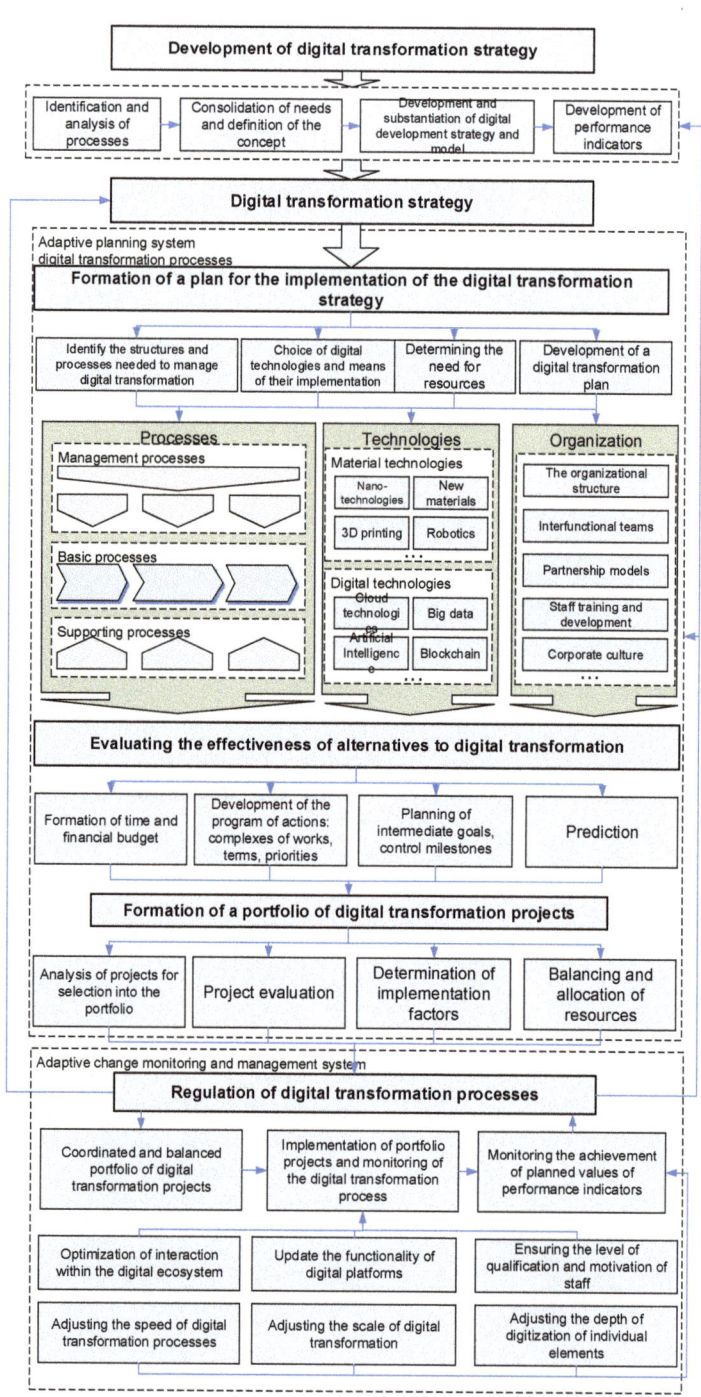

Figure 5. Multi-level model of digital transformation management in a high-tech enterprise. Source: own study.

When planning the process of creating distributed high-tech manufacturing in the context of digital transformation, it is necessary to take into account the virtualization of manufacturing, constraints associated with expensive land resources, and minimizing the cost of locating communications and technological nodes. For this purpose, it is necessary to optimize costs at the manufacturing location.

A key role in solving the problems of simulation of the management processes for the development of high-tech enterprises in the context of digital transformation is to build a system of predictive analytics.

Further, the formation and study of logistics interactions shall be carried out in a distributed enterprise in the context of digitalization and virtualization of production. For this purpose, it is necessary to model the logistics of the life cycle of high-tech products, taking into account qualitative and quantitative estimates of costs, time, and risks.

Particular attention shall be paid to the rational selection of suppliers in the distributed manufacturing of high-tech products and evaluation of dynamic processes of supply of components in long logistics chains of high-tech manufacturing in the context of globalization and digital transformation. Therefore, it is necessary to take into account the emergence of threats and justify risks when planning cargo transportation in a globalized economy.

Due to the fact that the level of digitalization of high-tech enterprises differs, the system of organizational and economic tools of digital transformation shall take into account the integration and resource capabilities of these entities and be as adaptive as possible.

3. Results

The basis for developing the strategy and portfolio of digital transformation projects of a high-tech enterprise is the result of an analysis of its operations and the level of its technical and technological readiness for digital transformation. This analysis allows for the directions of the digital transformation of a particular enterprise.

An important result in understanding the evolutionary development of digital models of corporate governance is also the need and possibility of gradual implementation of individual components and blocks of information and analytical system of the enterprise, taking into account the level of its digital maturity, strategic priorities, and digital ecosystem development. Currently, there is no universal strategy for the development of described digital models, so each enterprise shall independently determine the course of its development.

The analysis resulted in the key directions of digital transformation identified in the example of a high-tech instrumentation enterprise, which are presented in Table 1.

Table 1. Potential directions and projects of digital transformation of an instrumentation enterprise.

Area	Key Projects and Technologies
Production and technical area	Heavy-duty PLM system implementation project (including PDM/CAD/CAM/CAE subsystems) that supports all stages of product development MES system implementation project, with modules for core production operations, technical control, and inventory management Project for the implementation of an automated energy accounting system Launching a robotic line Cloud-based IoT system for real-time data control and process management Internal automated logistics through the introduction and optimization of robotic systems Creation of new structures of the logistics supply chain
Personnel and management area	Project for implementing a corporate management information system based on SAP S4/HANA, modules that provide analytics for business solutions Launching digital quality assessment modules Conducting corporate professional development and retraining programs Establishment of an engineering center and a competence transfer center

Table 1. Cont.

Area	Key Projects and Technologies
Innovation area	Data-driven production management based on digital duplicates of equipment and manufacturing processes Establishment of a center for additive technologies Conducting virtual tests using digital product models
Financial and investment area	Engagement of venture investors in projects, submission of projects to corporate accelerators Investments in R&D
Market area	Use of mass customization model for enterprise products Development of a scalable digital platform for interaction with suppliers and customers After-sales service of high-tech products

When choosing the direction of development, enterprise management is challenged with assessing the use of digital assets and the potential of the enterprise for manufacturing new products and building a new business model. The level of participation of the digital potential of an enterprise is assessed by the level of its components in the implementation of the main business processes for the manufacturing and sale of products. It is the main process that creates additional value to input resources and forms the final indicators of production.

Table 2 show a list of identified key digital assets of the instrumentation enterprise. Within our approach, the main factor in the success of the digital transformation is the active use of key digital assets and potential in the development of new markets, business models, and new product development. Those products, which involve key digital assets and potential in their manufacturing and sale, have strong competitive positions and prospects for further development.

Table 2. Key digital assets of an instrumentation enterprise.

Key Digital Assets of the Enterprise	Description	Availability during Digital Transformation of the Enterprise
Digital product profile management based on the PLM system	The enterprise has already implemented several subsystems of the PLM system. It provides full traceability throughout the product life cycle: from the design of individual components, including control at the manufacturing stage, to the operation of the finished product	Potential opportunity and basis for building a digital factory using digital product design technologies
Available manufacturing of unique customized products with great variability	Due to the presence of a large number of different machines and test benches. Extensive practice in the development and manufacture of a wide range of components and their systematic adjustment to the main equipment	Most precise customer order, which can be immediately sent to manufacturing, reducing preparation and production time
Cloud-based system for real-time data control and process management	IoT-based system that combines end devices and sensors and collects data from various automated production systems	Centralized management of accumulated data, and availability to connect analytics based on artificial intelligence and machine learning methods. In addition, it monitors operational processes in real-time, alerts about malfunctions in advance, performs intelligent diagnostics and decision-making

Table 2. Cont.

Key Digital Assets of the Enterprise	Description	Availability during Digital Transformation of the Enterprise
Unique technologies and production equipment, including CNC machines and robotic systems	Most of the equipment is unique. Production lines are equipped with robotic systems	With the available equipment, unique and competitive products can be manufactured to meet the specific needs
Stable financial position	The financial condition of the enterprise is significantly more favorable than the industry average	The resulting amount of profit and a stable financial position expand the possibilities of the enterprise for the development and introduction of new product directions, ensuring their financing in necessary volume and the shortest terms
Developed scientific base	The enterprise has a large number of patents and unique test benches	It is possible to organize the manufacturing of products that have unique properties and patent protection. Unique test benches ensure conducting field tests as close as possible to the working conditions, which in turn allows the enterprise to reach a high global level in terms of research, development, and production equipment in manufacturing instrumentation products
High-quality products	The enterprise is certified by the quality system of international standards and implements modules of technical control of the corporate governance system	Using the existing quality system and control methods allows the enterprise to create high-quality products that meet international standards
Available processing of orders for research and development projects from other enterprises in the region and industry	Ecosystem stakeholders place R&D orders on various topics	In the future, the enterprise can master mass manufacturing on the basis of R&D operations
Available processing of defense orders	In the current conditions in Ukraine, the enterprise can process state defense orders	Order availability creates medium-term and guaranteed sales, as well as further service for products
Highly qualified personnel	Workers and design specialists are the most qualified personnel in the enterprise	Highly qualified personnel provide an opportunity to create, develop and produce technologically complex products and further transition to Industry 4.0 technology
Extensive experience in foreign economic activity	The enterprise has been supplying products to foreign countries for many years	The enterprise has the knowledge and experience of working in international markets, which can be used to create and promote new products

Based on internal analysis and customer experience analysis, the following principal primary requirements for the properties of enterprise products were identified: low cost of ownership of products; high quality; reliability; application of innovative technologies; expanded range of services; complete delivery; minimum delivery time.

Further, the detailed structure of the consumer value of the enterprise products will help understand the internal components and proportions of the final properties of the product perceived by the consumer, capabilities, resources and digital assets, and the extent of involvement in the production of these components, which will help focus more precisely on the strategic plan for digital transformation.

Subsequently, the analysis of business processes was carried out, and the resources of the enterprise were analyzed. The method proposed in the paper allows the assessment of the participation of key digital assets of the enterprise in the implementation of principal business processes for the manufacturing and sale of products.

Table 3 show some of the results of calculations based on the developed method for an instrumentation enterprise. It shows expert scores on the use of digital assets of the enterprise in the implementation of major business processes in the principal product areas.

Table 3. Assessment of the use of digital assets in the implementation of major business processes of an instrumentation enterprise.

Product Areas	Major Business Processes	Digital Assets and Enterprise Potential											Total
		PLM System	Unique Technologies and Manufacturing Equipment	Stable Financial Position	Developed Scientific Base	Available Manufacturing of Unique Customized Products with Great Variability	Cloud-Based Data Management and Process Management System	High-Quality Products	Available Processing of Orders for Research and Development	Available Processing of Defense Orders	Highly Qualified Personnel	Experience in Foreign Economic Activity	
Product 1	Product development and modernization	4	0	3	4	3	2	0	2	2	4	0	24
	Procurement management	0	0	2	0	0	2	3	0	0	0	0	7
	Manufacturing	4	4	2	2	3	4	4	0	4	4	0	31
	Marketing management	0	0	0	0	0	1	0	0	0	0	0	1
	Installation, adjustment and testing	3	0	2	0	1	2	0	0	3	0	4	15
	Service	2	2	2	0	0	2	0	0	3	0	2	13
Product 2	Product development and modernization	4	0	1	2	4	2	0	3	2	3	0	21
	Procurement management	0	0	1	0	0	1	1	0	0	0	0	3
	Manufacturing	3	3	2	0	3	3	2	0	2	1	0	19
	Marketing management	0	0	0	0	0	1	0	0	0	0	0	1
	Installation, adjustment and testing	3	0	1	0	3	2	0	3	1	0	2	15
	Service	1	1	1	0	0	1	0	0	1	0	0	5
Product 3	Product development and modernization	4	0	1	3	3	4	0	3	0	3	0	21
	Procurement management	0	0	1	0	0	2	2	0	0	0	0	5
	Manufacturing	3	3	2	2	2	4	3	0	1	3	0	23
	Marketing management	0	0	1	0	0	2	0	0	0	0	0	3
	Installation, adjustment and testing	2	0	2	0	1	2	0	2	1	0	2	12
	Service	2	0	1	0	0	1	0	1	0	2	0	7

Processing the received data according to the offered procedure allows selection of those directions among all possible variants of digital transformation of the enterprise in which the maximum use of key digital assets and potential of the enterprise will be provided. Quantitative estimates of the degree of dependence or impact were formed on a five-point scale. A point score from 0 to 4 reflects the increasing degree of using a single key digital asset in the implementation of major business processes for the manufacturing and sale of enterprise products.

Therefore, the calculations showed that the key digital assets of the enterprise are most actively involved in the implementation of business processes for manufacturing, product development, and installation and commissioning (Figure 6).

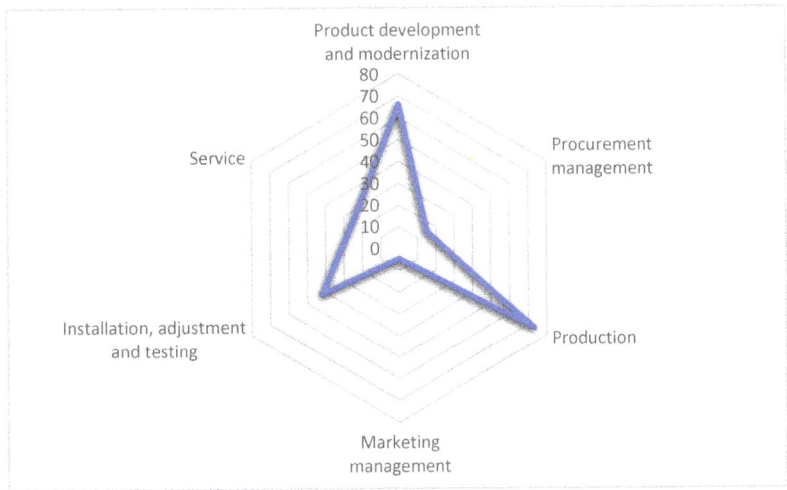

Figure 6. The level of use of digital assets of the enterprise in the implementation of major business processes.

Thus, several projects were selected for the digital transformation portfolio. The first is related to the ongoing implementation of the PLM system at the enterprise, which provides an automatic assessment of the technological implementation of manufacturing in the early stages of product design.

A digital product management platform has been developed and implemented, and it provides full traceability (storage and dissemination of technological and manufacturing data) throughout the product life cycle: from the design of individual parts and assemblies, including control at the production stage, to the operation of the finished product.

The next most time-consuming task is the MES system implementation project, which includes subsystems of operational manufacturing management, technical control (quality) and inventory management. Production planning based on these digital tools will help increase labor productivity and manufacturing rhythm and save resources, and the analysis of accumulated information will help identify reserves for reducing costs and improving technical processes.

The next project is related to the ongoing implementation of the IoT platform in the enterprise, which combines various end devices and high-speed communication channels, collects data from various automated manufacturing systems, carries out centralized management of accumulated data, and performs intellectual analysis and decision-making. In addition, it monitors operational processes in real-time, alerts about malfunctions in advance, and performs intelligent diagnostics and decision-making. Based on data on the operation of production units, their digital counterparts, which are the data source for the MES system, are created.

The next project is related to the use of a digital platform for customer service. A service business is a source of stable cash income, with possible options such as creating a service company from scratch or buying a player already operating in the market.

Thus, the proposed approach allows assessment of the degree of use of digital assets and the potential of the enterprise in the manufacturing and sale of certain products and provides the selection of the most promising areas of digital transformation that improve enterprise performance and competitiveness.

4. Discussion

The study results in the proposed methodology for managing the development of high-tech enterprises in the digital transformation based on the formation of an ecosystem model of decentralization in a single distributed digital space, based on interconnected adaptive systems of planning, monitoring, and change management, and, on the basis of modeling and forecasting of complex production and logistics processes of high-tech industries, it allows effective implementation of the innovative order portfolio in the short term and limited opportunities while coordinating the priorities of the business strategy and the strategy of digital transformation of high-tech enterprises.

A conceptual model of enterprise competitiveness formation in the process of digital transformation has been developed, which includes organizational and economical digital tools for sustainable development of high-tech enterprises and synergies from the organization of new forms of digital interaction. The system of the core elements of the mechanism of development of an industrial enterprise in the context of digitalization is described.

A multi-level model of digital transformation management in a high-tech enterprise has been developed based on decentralized mechanisms for allocating limited resources and predictive analytics for the effective implementation of an innovative order portfolio.

Currently, the practical implementation of the considered approaches and models is carried out primarily in terms of providing automated systems and the Internet of Things platform, predictive analytics modules, and designing intelligent decision support systems. In the short term, advanced analytics based on predictive modeling and machine learning will be developed for the creation of recommendation systems and implemented in the form of cloud platforms and services.

In conclusion, it should be noted that the development of digital models of corporate governance of each enterprise involves systematic work on the allocation of priority digital assets in the information system of the organization, as well as overcoming organizational, managerial and technological obstacles.

Author Contributions: Conceptualization, Y.P., O.P. and O.F.; methodology Y.P., O.P. and O.F.; writing—original draft preparation, Y.P., O.P. and O.F.; writing—review and editing, Y.P., O.P. and O.F.; supervision, Y.P., O.P. and O.F. All authors have read and agreed to the published version of the manuscript.

Funding: This research received no external funding.

Institutional Review Board Statement: Not applicable.

Informed Consent Statement: Not applicable.

Data Availability Statement: Not applicable.

Conflicts of Interest: The authors declare no conflict of interest.

References

1. Machado, C.; Winroth, M.; Carlsson, D.; Almström, P.; Centerholt, V.; Malin, H. Industry 4.0 Readiness in Manufacturing Companies: Challenges and Enablers towards Increased Digitalization. *Procedia CIRP* **2019**, *81*, 1113–1118. [CrossRef]
2. Ghobakhloo, M. Industry 4.0, Digitization, and Opportunities for Sustainability. *J. Clean. Prod.* **2020**, *252*, 119869. [CrossRef]
3. Cimini, C.; Pezzotta, G.; Pinto, R.; Cavalieri, S. Industry 4.0 Technologies Impacts in the Manufacturing and Supply Chain Landscape: An Overview. In *Service Orientation in Holonic and Multi-Agent Manufacturing*; Borangiu, T., Trentesaux, D., Thomas, A., Cavalieri, S., Eds.; Springer: Berlin/Heidelberg, Germany, 2019. [CrossRef]

4. Balakrishnan, R.; Das, S. How do firms reorganize to implement digital transformation? *Strateg. Chang.* **2020**, *29*, 531–541. [CrossRef]
5. Smuts, S.; van der Merwe, A.; Smuts, H. A Strategic Organisational Perspective of Industry 4.0: A Conceptual Model. In *Responsible Design, Implementation and Use of Information and Communication Technology*; Hattingh, M., Matthee, M., Smuts, H., Pappas, I., Dwivedi, Y.K., Mäntymäki, M., Eds.; Springer: Berlin/Heidelberg, Germany, 2020; Volume 12066, pp. 89–101. [CrossRef]
6. Kushnirenko, O.M. Industry of Ukraine facing Industry 4.0 challenges: Assessment of constraints and policy objectives. *Econ. Ukr.* **2020**, *5*, 53–71. [CrossRef]
7. Digital Business Transformation. A Conceptual Framework. 2015 Global Center for Digital Business Transformation. Available online: https://ru.scribd.com/document/372049639/Digital-Business-Transformation-Framework.pdf (accessed on 20 May 2022).
8. Schallmo, D.; Williams, C.A.; Boardman, L. Digital transformation of business models–best practice, enablers and roadmap. *Int. J. Innov. Manag.* **2017**, *21*, 1740014. Available online: https://www.worldscientific.com/doi/pdf/10.1142/S136391961740014X (accessed on 29 May 2022). [CrossRef]
9. Sjodin, D.R.; Parida, V.; Leksell, M.; Petrovic, A. Smart Factory Implementation and Process Innovation: A Preliminary Maturity Model for Leveraging Digitalization in Manufacturing. *Res. Technol. Manag.* **2018**, *61*, 22–31. [CrossRef]
10. Parida, V.A.; Sjödin, D.R.; Wincent, J.; Kohtamäki, M. Survey study of the transitioning towards high-value industrial product-services. *Procedia CIRP* **2014**, *16*, 176–180. [CrossRef]
11. Zemlickienė, V. Customization Model for Assessing the Commercial Potential of Technologies to Different Technology Manufacturing Branches: Development a Set of Factors. *Proceedings* **2018**, *2*, 1522. [CrossRef]
12. Schumacher, A.; Nemetha, T.; Sihna, W. Roadmapping towards industrial digitalization based on an Industry 4.0 maturity model for manufacturing enterprises. In Proceedings of the 12th CIRP Conference on Intelligent Computation in Manufacturing Engineering, Gulf of Naples, Italy, 18–20 July 2018; Volume 79, pp. 409–414. [CrossRef]
13. Teng, X.; Wu, Z.; Yang, F. Research on the Relationship between Digital Transformation and Performance of SMEs. *Sustainability* **2022**, *14*, 6012. [CrossRef]
14. Bressanelli, G.; Adrodegari, F.; Pigosso, D.C.A.; Parida, V. Towards the Smart Circular Economy Paradigm: A Definition, Conceptualization, and Research Agenda. *Sustainability* **2022**, *14*, 4960. [CrossRef]
15. Castelo-Branco, I.; Cruz-Jesus, F.; Oliveira, T. Assessing Industry 4.0 readiness in manufacturing: Evidence for the European Union. *Comput. Ind.* **2019**, *107*, 22–32. [CrossRef]
16. Santos, R.C.; Martinho, J.L. An Industry 4.0 maturity model proposal. *J. Manuf. Technol. Manag.* **2019**, *31*, 1023–1043. [CrossRef]
17. Li, J.; Zhou, J.; Cheng, Y. Conceptual Method and Empirical Practice of Building Digital Capability of Industrial Enterprises in the Digital Age. *IEEE Trans. Eng. Manag.* **2019**, *69*, 1902–1916. [CrossRef]
18. Warner, K.S.R.; Wäger, M. Building dynamic capabilities for digital transformation: An ongoing process of strategic renewal. *Long Range Plan.* **2019**, *52*, 326–349. [CrossRef]
19. Butt, J. A Strategic Roadmap for the Manufacturing Industry to Implement Industry 4.0. *Designs* **2020**, *4*, 11. [CrossRef]
20. Parida, V.; Sjödin, D.; Reim, W. Reviewing Literature on Digitalization, Business Model Innovation, and Sustainable Industry: Past Achievements and Future Promises. *Sustainability* **2019**, *11*, 391. [CrossRef]
21. Kraus, S.; Durst, S.; Ferreira, J.; Veiga, P.; Kailer, N.; Weinmann, A. Digital transformation in business and management research: An overview of the current status quo. *Int. J. Inf. Manag.* **2021**, *63*, 102466. [CrossRef]
22. Prokhorov, O.; Pronchakov, Y.; Prokhorov, V. Cloud IoT Platform for Creating Intelligent Industrial Automation Systems. *Lect. Notes Netw. Syst.* **2021**, *188*, 55–67.
23. Zhong, R.Y.; Xu, X.; Klotz, E.; Newman, S. Intelligent Manufacturing in the Context of Industry 4.0: A Review. *Engineering* **2017**, *3*, 616–630. [CrossRef]
24. Chang, V.I.; Lin, W. How Big Data Transforms Manufacturing Industry: A Review Paper. *Int. J. Strateg. Eng.* **2019**, *2*, 39–51. [CrossRef]
25. Gürdür, D.; El-khoury, J.; Törngren, M. Digitalizing Swedish industry: What is next? *Comput. Ind.* **2019**, *105*, 153–163. [CrossRef]
26. Fedorovich, O.; Uruskiy, O.; Pronchakov, Y.; Lukhanin, M. Method and information technology to research the component architecture of products to justify investments of high-tech enterprise. *Radioelectron. Comput. Syst.* **2021**, *1*, 150–157. [CrossRef]
27. Prokhorov, O.; Pronchakov, Y.; Fedorovich, O.; Kunanets, N. Modeling of Technological Process in Nanoelectronic Production. In Proceedings of the 2020 IEEE 15th International Conference on Computer Sciences and Information Technologies (CSIT), Zbarazh, Ukraine, 23–26 September 2020; pp. 324–327. [CrossRef]
28. Romanenkov, Y.; Pronchakov, Y.; Zieiniiev, T. Algorithmic Support for Auto-modes of adaptive short-term Forecasting in predictive Analytics Systems. In Proceedings of the 2020 IEEE 15th International Conference on Computer Sciences and Information Technologies (CSIT), Zbarazh, Ukraine, 23–26 September 2020; pp. 230–233. [CrossRef]
29. Hallstedt, S.I.; Isaksson, O.; Öhrwall Rönnbäck, A. The Need for New Product Development Capabilities from Digitalization, Sustainability, and Servitization Trends. *Sustainability* **2020**, *12*, 10222. [CrossRef]

Article

Investigation of Statistical Machine Learning Models for COVID-19 Epidemic Process Simulation: Random Forest, K-Nearest Neighbors, Gradient Boosting

Dmytro Chumachenko [1,*], Ievgen Meniailov [1], Kseniia Bazilevych [1], Tetyana Chumachenko [2] and Sergey Yakovlev [1,3]

1 Mathematical Modelling and Artificial Intelligence Department, National Aerospace University "Kharkiv Aviation Institute", 71072 Kharkiv, Ukraine; evgenii.menyailov@gmail.com (I.M.); ksenia.bazilevich@gmail.com (K.B.); svsyak7@gmail.com (S.Y.)
2 Epidemiology Department, Kharkiv National Medical University, 61000 Kharkiv, Ukraine; tatalchum@gmail.com
3 Institute of Information Technology, Lodz University of Technology, 90-924 Lodz, Poland
* Correspondence: dichumachenko@gmail.com

Abstract: COVID-19 has become the largest pandemic in recent history to sweep the world. This study is devoted to developing and investigating three models of the COVID-19 epidemic process based on statistical machine learning and the evaluation of the results of their forecasting. The models developed are based on Random Forest, K-Nearest Neighbors, and Gradient Boosting methods. The models were studied for the adequacy and accuracy of predictive incidence for 3, 7, 10, 14, 21, and 30 days. The study used data on new cases of COVID-19 in Germany, Japan, South Korea, and Ukraine. These countries are selected because they have different dynamics of the COVID-19 epidemic process, and their governments have applied various control measures to contain the pandemic. The simulation results showed sufficient accuracy for practical use in the K-Nearest Neighbors and Gradient Boosting models. Public health agencies can use the models and their predictions to address various pandemic containment challenges. Such challenges are investigated depending on the duration of the constructed forecast.

Keywords: epidemic model; epidemic process; machine learning; COVID-19; K-Nearest Neighbors method; gradient boosting; random forest

1. Introduction

The COVID-19 virus was first reported in December 2019 [1]. Chinese authorities told the World Health Organization (WHO) that a man died from a respiratory disease of unknown origin in Wuhan, Hubei province. In early January 2020, it was revealed that the genome of a new type of coronavirus is similar to the genome of the SARS virus that spread worldwide from China in 2002–2003 [2]. Initially, the new coronavirus was treated by the world health system as an epidemic of a regional scale, affecting only China. Nevertheless, in the first month, the virus began to spread rapidly outside of China and threatened the health of the entire planet's population [3]. On 11 March 2020, the WHO declared a global pandemic of COVID-19.

The entire world community has directed efforts to prevent and combat the new coronavirus. The virus has spread across the globe through tourists and the availability of flights. In the spring of 2020, restrictive measures were introduced in most countries to contain the spread of COVID-19 [4]. Among such activities were lockdowns, contact tracing with isolation of contact individuals, the introduction of a mask regime, social distancing, etc. As the epidemic was contained, the authorities of individual countries began to gradually ease lockdowns and other restrictive measures to minimize damage to the economy and prevent social problems [5]. In the fall of 2020, the second and third waves of the epidemic began in many countries [6]. New strains of the virus began to spread, characterized by increased virulence [7].

A large-scale vaccination campaign, which was launched in a short time around the world, contributed significantly to the fight against COVID-19 [8]. The development of vaccines against coronavirus diseases, such as SARS and MERS, which began even before the onset of the COVID-19 pandemic, made it possible to form knowledge about the structure and rules of the coronaviruses spread [9]. Furthermore, it is this knowledge that has accelerated the development of various types of vaccines during the current pandemic. Many countries have introduced phased population vaccination plans, identifying groups at the highest risk of complications. Inactivated vaccines, live attenuated vaccines, vector non-replicating and vector replicating vaccines, vector inactivated, DNA and RNA vaccines, and recombinant protein vaccines have been developed. Some of them were used to combat the pandemic.

The unprecedented crisis caused by the global COVID-19 pandemic has demonstrated the significant role of digital technologies [10]. Since the beginning of the pandemic, the world has seen an accelerated digitalization of many activities, such as the economy [11], finance [12], business [13], transport [14], education [15], and many others. Digitalization has not bypassed the field of medicine with the improvement of diagnostic methods [16], automated processing of medical data [17], and storage of medical data [18]. Models and methods for modeling epidemic morbidity received a new round.

This study **aims** to develop three models for predicting the dynamics of the COVID-19 epidemic process in specific areas using statistical machine learning methods and to study the results of the experiments of the constructed models.

To achieve this goal, the following **tasks** were formulated:

1. To analyze models and methods for modeling the epidemic process of COVID-19.
2. To analyze data on the incidence of COVID-19 in the selected territories.
3. To develop a model for predicting the dynamics of the COVID-19 epidemic process based on the K-Nearest Neighbors method.
4. To develop a model for predicting the dynamics of the COVID-19 epidemic process based on Gradient Boosting.
5. To develop a model for predicting the dynamics of the COVID-19 epidemic process based on the Random Forest method.
6. To evaluate the results of an experimental study using the developed models.
7. To analyze the developed models for accuracy and computational complexity.

The promising **contribution** of this study is two-stage. First, the development of models based on statistical machine learning methods will make it possible to assess the accuracy of forecasts of the dynamics of the COVID-19 epidemic process built using simple models. Secondly, a comparative study of three models of statistical machine learning will allow us to conclude which of them is more effective for studying the epidemic processes not only of COVID-19 but also of other infectious diseases.

The further **structure** of the paper is the following: Section 2, Current Research Analysis, provides an overview of models and methods of epidemic process simulation. Section 3, Data on COVID-19 Morbidity Analysis, provides a brief description of the COVID-19 pandemic in countries investigated within the research: Germany, Japan, South Korea, and Ukraine. Section 4, Model and Methods, describes three regression approaches to COVID-19 morbidity forecasting. Section 4, Results, describes the results of models' performance, estimation of developed models' adequacy, and forecasting accuracy. Section 5, Discussion, discusses the perspective use of models and their limitations. The conclusion describes the outcomes of the research.

Research is part of a complex intelligent information system for epidemiological diagnostics, the concept of which is discussed in [19].

2. Current Research Analysis

The field of modeling epidemic processes originated at the beginning of the 20th century with the works of Ronald Ross [20], William Hamer [21], Anderson McKendrick, and William Kermack [22]. The works of these scientists laid the mathematical foundations of epidemiology, proposing to describe the dynamics of morbidity using compartmental models [23]. In such models, the population is divided into compartments depending on their belonging to a defined state. The epidemic process occurring in the population is described using systems of differential equations.

Compartment models are used to model and study many infectious diseases. The paper [24] describes the application of compartmental models to study measles incidence. Double vaccination was considered, and the model was studied for balance and stability. The results show that the rate of transmission of infection has the most significant impact on the incidence of measles. In [25], the incidence of influenza was considered, and the classical SIR model was studied. The application of the probabilistic approach in the transition between states is considered. It is concluded that the negative aspect of applying the compartmental approach to modeling influenza is the non-obviousness of the results concerning one or even several scenarios of the development of the epidemic. The compartmental approach to influenza modeling was used as early as the 1970s by Baroyan and Rvachev [26]. The simulation results were used in the USSR to substantiate anti-epidemic measures aimed at combating the increase in the incidence of influenza.

Among intestinal infections, a compartmental approach is applied to modeling salmonellosis. The study [27] considered non-infectious and endemic resistant states. The model itself is not accurate enough to conduct relevant experiments to study the dynamics of Salmonella bacterial infection. The model of the hepatitis A built-in [28] aims to assess the impact of various vaccination strategies. The results show the importance of hepatitis A vaccination in early childhood.

In [29], an air-borne infection diphtheria incidence model was constructed by extending the classical SIR model. The authors found a globally asymptotically stable equilibrium of infectious extinction. However, such results cannot be effectively interpreted in epidemiology and public health.

The compartmental approach also applies to infections with a contact route of transmission. The work [30] is devoted to modeling HIV/AIDS with the possibility of treatment. The authors have proved that painless equilibrium is globally asymptotically stable when the base reproduction number is less than one. However, such a conclusion is a law of epidemiology and does not require analytical proof using modeling tools. The model of hepatitis B described in [31] shows the importance of assessing population migration for the spread of the disease. The authors claim that it is possible to reduce the incidence of hepatitis B based on the model results. However, the main vectors of the infection are not taken into account when compiling compartments. The authors of [32] describe a model of the epidemic process of hepatitis C. The emphasis is on people who inject drugs. The model is dynamic and interactively presented using a web application. The disadvantage of the model is that if there is a significant change in the rules of distribution, for example, the introduction of a policy to combat injecting drug users or the introduction of mass substitution therapy, all model parameters must be adjusted again.

A common disadvantage of the models described above is the impossibility of extending them to other objects. It is necessary to completely rebuild the model and find new coefficients related to a particular disease to model another disease.

With the onset of the global COVID-19 pandemic, compartmental models are actively used to model the epidemic process of a new coronavirus in various territories. Such territories can have different sizes, densities, and populations. Thus, in [33], the territory of the college campus is considered, where complex public health protocols can be introduced. In [34], the spread of COVID-19 in New York is modeled to determine the peak of the incidence wave. The work [35] extends the territory of modeling to the state of New York. It examines strategies to manage the course of the epidemic based on control measures implemented in other states. In [36,37], the dynamics of COVID-19 are modeled on the island states, limited from the outside world of Sri Lanka and Cyprus. In the case of Sri Lanka, the emphasis is on the isolation of villages on the island and the absence of tourists. When modeling the epidemic situation in Cyprus, an arbitrary number of subgroups with different infection levels and testing were used. In [38], an entire country was taken to model COVID-19: France. New cases, deaths, hospitalizations, intensive care unit admissions, hospital deaths, etc., are used. In [39], several European countries are considered at once, and for each country, its transmission coefficients, recovery rates, etc., are calculated. Considering compartmental models for different areas, it should be noted that even when studying a single disease, such as COVID-19, the coefficients of the model should be found again for each area, and the system of differential equations should be rebuilt from the very beginning.

Compartmental approaches with different sets of states are also used to model COVID-19. The study [40] uses the simplest SIR (susceptible—infected—recovered) model. The disadvantage of the model is the accuracy of forecasts, which is insufficient for decision-making, and the limitedness in population groups gives a very general understanding of the spread of the epidemic process. In [41], the classical SIR model is extended by adding the exposed state. The model is used to find the peak of the disease, but the results have not materialized due to changes in the policy of control measures in the countries considered and the start of the vaccine campaign. The work [42] extends the classical SIR model with the state Q—quarantined for isolated infected people. The model shows that the maximum number of infected in the real world is highly dependent on the speed with which quarantine restrictions are implemented. The authors of [43] add the D-death state to the SEIR model for fatal cases. Modeling results show that unreported deaths from COVID-19 are significantly lower than unreported infections. In [44], the authors extend the SEIR model with the state Q—quarantined. At the same time, the model does not consider isolation scenarios and social distancing. The study [45] extends the SEIR model with states D—death and Q—quarantined. Moreover, the quarantined state means hospitalization since the authors hypothesize that hospitalization is similar to quarantine restrictions. In this case, the model considers the average behavior of the population, which leads to an underestimation of specific population groups. In [46], the authors present a model consisting of seven compartments: susceptible (S), exposed (E), infectious (I), quarantined (Q), recovered (R), deaths (D), and vaccinated (V). The model can estimate predicted numbers of compartments, but only for a short time. Models with a much larger number of compartments are also known. However, a common disadvantage is that many states and subpopulations are needed to adequately describe a population, which makes models complex. The complexity of the models causes both difficulties with calculations and experimental studies and the impossibility of promptly making changes to the model when the behavior of the virus dynamics changes.

Models are also used for various tasks in the study of COVID-19. For example, work [47] considers the effectiveness of vaccination distribution. The study [48] looks at the transport effects of the COVID-19 pandemic. In [49], the effectiveness of the introduction of lockdowns is estimated. Ref. [50] explores the effects of social distancing. The authors of [51] investigate the effectiveness of masks to combat the novel coronavirus pandemic. The study [52] is devoted to assessing the economic aspects of applying control measures to combat the COVID-19 pandemic. Work [53] uses compartmental models to investigate the transmission of the COVID-19 virus among medical personnel and methods for protecting healthcare workers from infection. Ref. [54] uses modeling to estimate the medical throughput of hospitalization, including for intensive care units.

However, the compartmental approach to modeling infectious diseases, including COVID-19, has several disadvantages, among which are the following:

- An accurate description of the population in which the epidemic process spreads requires considering the population's heterogeneity, i.e., age, gender, behavior, physical

- interaction, etc. However, introducing all these characteristics into the compartmental model significantly complicates it and makes it unsuitable for practical use.
- The apparatus of differential equations has high computational complexity with sufficiently detailed models.
- Different diseases have different conditions and rules of infection transmission in different population groups, making it impossible to transfer an already ready model for one infectious disease to another disease. So, for each new disease, the model must be rebuilt.
- The same model cannot be applied in different territories even for the same disease because transfer rules and control measures may differ depending on the location, climate, legal aspects, etc. For each new territory, the model needs to be built anew.
- When the virulence of the disease changes, it is impossible to make changes to the model quickly, and all coefficients must be re-found experimentally. The rate at which model changes are made is especially critical when modeling COVID-19, as the virus mutates rapidly and new strains have different dynamics while circulating in the population along with known strains.
- The non-adaptation of compartmental models to external factors makes it impossible to predict for medium and long-term periods. Sufficient accuracy for studying the epidemic process can be obtained only when calculating a short-term forecast.

Based on the analysis, we will use statistical machine learning models to eliminate the shortcomings of compartmental models. Such models are characterized by high predictive accuracy, adaptability, and the ability to overtrain models during a pandemic based on updated data, the ability to use a comprehensive set of population data to display more realistic behavior of the virus.

3. Data on COVID-19 Morbidity Analysis

Data on new cases of COVID-19 aggregated by the Johns Hopkins University Coronavirus Resource Center was used for the experimental study [55]. Data on the incidence of COVID-19 in Germany, Japan, South Korea, and Ukraine were selected for analysis. These countries were chosen because the dynamics of the pandemic were different, and the decision-makers implemented different anti-epidemic measures to curb the incidence. The different nature of the pandemic makes it possible to verify the constructed models and evaluate their accuracy and adequacy on different samples.

3.1. COVID-19 in Germany

Since the beginning of the pandemic, data in Germany have been recorded and analyzed by the Robert Koch Institute [56]. As of April 2022, more than 23.5 million cases have been registered in Germany, of which more than 130 thousand are fatal. The first case of COVID-19 in Germany was reported on 27 January 2020. On 17 March 2020, schools and kindergartens were closed in all federal states of Germany, and a state of emergency was introduced in Bavaria. On 25 March 2020, the Bundestag declared an epidemic situation of national importance [57]. Since May 2020, some restrictions have been lifted and tightened again in October 2020. Since December 2020, a national lockdown had been introduced, which extended lately till the beginning of March 2021. In May 2021, two counties reported no cases of COVID-19 for the first time.

Furthermore, until July 2021, the incidence in Germany was declining. The further development of the pandemic in Germany is associated with the emergence of the delta strain. The growth continued until December 2021 [58]. At the beginning of 2022, the pandemic in Germany was characterized by the widespread Omicron strain. Since January 2022, there has been an increase in incidence. At the same time, a significant increase in mortality rates is not observed [59].

As of April 2022, there have been five waves of COVID-19 in Germany. The first wave (March–April 2020) and the second wave (October 2020–January 2021) are characterized by a disproportionate impact on older populations, resulting in a high number of deaths. It should be noted that the number of deaths in Germany was still lower than in other countries. This is due to the excellent equipment of German hospitals. The share of intensive care beds in the population is one of the highest in the world [60]. However, the number of occupied beds in intensive care units also increased during the third and fourth waves. The availability of vaccines since December 2020 has reduced mortality since the third wave. The pandemic in Germany is characterized by the spread of the alpha strain from March 2021, the delta strain from June 2021, and the Omicron strain from January 2022.

The COVID-19 vaccination rate as of April 2022 is 75.08% of those who received the entire course of vaccination. 58.33% of the population received a booster dose of the vaccine [61].

3.2. COVID-19 in Japan

As of April 2022, almost 7.5 million cases of COVID-19 were registered in Japan, of which almost 30 thousand were fatal. The first case of COVID-19 in Japan was registered on 16 January 2020, by a citizen who arrived from Wuhan (China). The following outbreaks were due to travel from Europe and the United States in March 2020 [62]. At the same time, strains characteristic of the European region prevailed in the country, and the Wuhan strain disappeared in March 2020. In February 2020, all primary, incomplete and secondary schools were temporarily closed. In April 2020, a state of emergency was declared. Despite the high prevalence of the virus, the mortality rate in Japan is one of the lowest [63]. This is due to the high level of mandatory testing of the population. In addition to testing, this was also influenced by the cultural habits of citizens, such as bow etiquette, wearing face masks, washing hands with disinfectants, etc. In the summer of 2021, the Olympic Games took place in Japan, which entailed numerous new restrictions to avoid new virus outbreaks.

The pandemic in Japan can be divided into five waves. The first wave was characterized by the Wuhan strain, which predominated in patients from China and other East Asian countries. The second wave was characterized by variants of the European type, which came to Japan with travelers from Europe [64].

In addition to containing the virus, government efforts have also focused on strengthening the health care system. This made it possible to strengthen the system of testing and consultation of patients in the hospital system. Special counseling centers and outpatient departments were established in medical institutions [65]. General health facilities in areas of COVID-19 outbreaks have been accepting patients with suspected infection. Additionally, the new medical policy allowed people with symptoms not to go to work and isolate themselves at home. Moreover, an effective contact tracing system has been developed since February 2020 to contain the spread of the virus.

The vaccination rate against COVID-19 as of April 2022 in Japan is 80.47% of those who received the entire vaccination course. 49.61% of the population received a booster dose of the vaccine [66].

3.3. COVID-19 in South Korea

In South Korea, as of April 2022, more than 16.5 million people fell ill, of which more than 21.5 thousand cases were fatal. The first case of COVID-19 in South Korea was registered on 20 January 2020. For the first four weeks, South Korea controlled the potential spread of the virus. For this, high-tech tools were used, including tracking the use of credit cards, analyzing CCTV footage of infected patients, and so on [67]. However, in mid-February, an outbreak of the disease nevertheless arose due to the infection of a member of a religious sect. Through the members of the sect, the disease spread rapidly. By March 2020, church-related infections accounted for 62.8% of all cases [68]. The government implemented immediate control measures, such as isolating patients, closing places where the infection was detected, and others. With the increase in the incidence in other countries, restrictive measures related to entry into the country were introduced in April 2020.

Measures to contain the pandemic in South Korea are considered the most effective in the world [69]. They included mass testing of the population for the virus, isolation of all patients, and tracking and isolation of all contact people. The rapid and extensive testing that the South Korean health system has been able to carry out has successfully limited the outbreak's spread without resorting to many area quarantine restrictions [70]. In South Korea, there was no general lockdown of businesses. Shops and supermarkets were open. Additionally, depending on the dynamics of the spread of the virus, the levels of distancing of the population have been introduced. Since February 2021, a large-scale vaccination campaign has begun. In January 2022, as in the rest of the world, the number of cases increased with the new Omicron strain. However, by early March 2022, South Korea began to relax social distancing rules, and by March 18, it had moved to an endemic lifestyle.

The vaccination rate against COVID-19 as of April 2022 in South Korea is 86.72% of those who completed the entire vaccination course. 64.31% received a booster dose [71].

3.4. COVID-19 in Ukraine

In Ukraine, as of April 2022, almost 5 million people had fallen ill, of which more than 100 thousand cases have been fatal. The first case was registered on 3 March 2020, by a citizen who returned from Europe. Even before the registration of the first case, a decision was made to conduct temperature screening of all citizens who stay in Ukraine. Since 12 March 2020, quarantine has been introduced in the country, including the closure of educational institutions, and holding public events. From 16 March 2020, the borders were closed to foreigners. Later, movement by public transport was limited, and the subway was closed. Since 25 March 2020, a state of emergency has been introduced throughout Ukraine [72]. Since April 2020, the Diy Vdoma mobile application has been introduced to track isolation for citizens who need mandatory isolation or observation. By the summer of 2020, quarantine restrictions were eased in several stages. Since July 2020, an adaptive quarantine has been introduced, assigning a quarantine zone to a region depending on the incidence rates. So, the corresponding restrictions apply in the region depending on belonging to the zone (green, yellow, orange, red) [73].

In February 2021, a vaccination campaign began in Ukraine. Since June 2021, vaccination has become available to all population categories. However, the percentage of vaccinated citizens has remained low. This is due both to the lack of confidence in the vaccination of some population groups and to the active anti-vaccination campaign on the part of Russia [74]. In October 2021, the government introduced mandatory vaccination of specific population groups, including teachers and education workers, civil servants, and medical workers. The vaccination rate against COVID-19 as of April 2022 is 36.96% of the population. According to the data, 1.76% of the population has received a booster dose [75].

Since January 2022, there has been an increase in the incidence associated with the Omicron strain. Since 24 February 2022, the incidence registration has become more complicated due to the Russian military invasion of Ukraine. The data is limited to severe cases, and registration is not possible in areas with active hostilities and in temporarily occupied territories. Therefore, the data on the incidence of COVID-19 in Ukraine, which are used in this study, include data up to 24 February 2022, and do not include data from the territories of Donetsk, Luhansk regions, and Crimea temporarily occupied by Russia.

4. Models and Methods

As part of this study, three models for predicting new cases of COVID-19 were built based on regression methods. The models are based on the Random Forest, K-Nearest Neighbors regression, and Gradient Boosting methods.

Regression analysis is a set of statistical methods for assessing the relationship between variables [76]. It can be used to model future relationships between variables, i.e., forecasting. Regression shows how changes in independent variables can be used to fix changes in dependent variables. In our case, the independent variables are the incidence of COVID-19, and the dependent variables are the predicted incidence.

4.1. Random Forest Model

A Random Forest is a machine learning algorithm that consists of many decision trees [77]. It uses bootstrap and feature randomness to build each individual tree to create an uncorrelated forest that has a better prediction than any individual tree.

The algorithm for constructing a Random Forest consisting of N trees can be represented as follows:

For every $n = 1, \ldots, N$:

- Generate sample X_n using bootstrap.
- Construct a decision tree b_n by the sample X_n.
- According to the given criterion, choose the best attribute, do a split in the tree according to it, and do it until the sample is exhausted.
- The tree is built until there are no more than n_{min} objects in each leaf or until a certain height of the tree is reached.
- For each partition, select m random features from n initial ones to find the optimal separation among them.

The final regression algorithm looks like this:

$$f(x) = \frac{1}{N} \sum_{i=1}^{N} b_i(x), \quad (1)$$

where $b_i(x)$ is a regression tree.

The recommended number of random features in regression tasks is $m = n/3$, where n is the number of initial features.

To improve the accuracy of forecasting by the Random Forest method, it is necessary to:

- Have features that have some predictive power.
- Uncorrelated forest tree predictions.
- Correct choice of features and hyperparameters for constructing weak correlations.

The random subspace method reduces the correlation between trees and avoids overfitting. The basic algorithm is trained on various subsets of the feature description, which are selected randomly. The ensemble of models using the random subspace method has the following construction algorithm:

- Let the number of objects for learning be N, and the number of features D.
- Choosing the number of individual models L in the ensemble is necessary.
- For each individual model l, it is necessary to choose dl ($dl < D$) as the number of features for l.
- It is necessary for each individual model l to create a training sample by selecting dl features from D and to train the model.
- It is necessary to combine the results of individual L models by combining the posterior probabilities.

4.2. K-Nearest Neighbors Model

The K-Nearest Neighbors method is a machine learning method based on finding the nearest objects with known target variable values [78]. For the regression problem, the average method is usually used, and the forecasting result is the average value of the last K sample data.

To build a model, a training sample is required, on which the correspondence "group of objects"—"dependent variable" is set:

$$X^m = (x_1, y_1), \ldots, (x_m, y_m) \quad (2)$$

The distance function between objects must be uniquely specified on the set of objects. For a random object, the method determines the distance to objects of a particular class and arranges them in ascending order:

$$p(u, x_{1,u}) \leq p(u, x_{2,u}) \leq \ldots \leq p(u, x_{m,u}) \tag{3}$$

where $x_{i,u}$ is the i-th neighbor of object u,
$y_{i,u}$ is the i-th neighbor for the dependent variable.
In general, the regression function looks like this:

$$\hat{y} = \frac{\sum_{k=1}^{K} y_k}{K} \tag{4}$$

where K is selected by cross-validation, and the metric is selected based on the selected feature space.

In this case, the class boundaries will be very complex, which contradicts that the method has one parameter. However, the paradox is resolved by the fact that the objects of the training sample are also peculiar parameters of the method.

Cross-validation evaluates an analytical model and its behavior on independent data, using the available data as evenly as possible.

Advantages of the method:

- Knowledge of features is optional, and only the proximity function is needed.
- The method applies to objects of any complexity if the proximity function is specified.
- Easy to implement.
- Easy to interpret.

The disadvantage of the method is that the accuracy of the method deteriorates with increasing space dimension.

4.3. Gradient Boosting Model

Gradient Boosting is a machine learning technique for classification and regression problems that builds a prediction model in an ensemble of weak predictive models [79]. In our case, Gradient Boosting is an ensemble of decision trees. The method is based on iterative learning of decision trees to minimize the loss function. Thanks to the features of decision trees, Gradient Boosting can work with categorical features and cope with non-linearities. Boosting is a method of transforming poorly trained models into well-trained ones. In boosting, each new tree is trained on a modified version of the original dataset.

Let there be a set of pairs of features x and target variables y, $\{(x_i, y_i)\}_{i=1,\ldots,n}$, on which it is necessary to restore the dependence of the form $y = f(x)$. It is necessary to minimize the loss function $L(y,f)$, which must be differentiable:

$$y \approx \hat{f}(x) \tag{5}$$

$$\hat{f}(x) = \underset{f(x)}{\operatorname{argmin}} L(y, f(x)) \tag{6}$$

It is necessary to find approximations $\hat{f}(x)$ in such a way as to minimize the loss function on the average on the available data. We restrict the search space to a parameterized family of functions $f(x, \theta)$, $\theta \epsilon R^d$. Then the problem is reduced to the one solved by optimizing the parameter values:

$$\hat{f}(x) = f(x, \hat{\theta}) \tag{7}$$

$$\hat{\theta}(x) = \underset{\theta}{\operatorname{argmin}} E_{x,y}(L(y, f(x, \theta))) \tag{8}$$

Find the approximate value of the parameters iteratively.

$$\hat{\theta} = \sum_{i=1}^{M} \hat{\theta}_i \qquad (9)$$

$$L_\theta(\hat{\theta}) = \sum_{i=1}^{N} L(y_i, f(x_i, \hat{\theta})) \qquad (10)$$

where $L_\theta(\hat{\theta})$ is empirical loss function, M is number of iterations.

To minimize $L_\theta(\hat{\theta})$ using the gradient descent method. To do this, it is necessary to initialize the initial approximation of the parameters $\hat{\theta} = \hat{\theta}_0$. For each iteration $t = 1, \ldots, M$, the following steps must be performed:

To calculate the gradient of the loss function $\nabla L_\theta(\hat{\theta})$ at the current approximation $\hat{\theta}$

$$\nabla L_\theta(\hat{\theta}) = \left(\frac{\partial L(y, f(x, \theta))}{\partial \theta} \right)_{\theta = \hat{\theta}} \qquad (11)$$

To set the current iterative approximation $\hat{\theta}_t$ based on the computed gradient.

$$\hat{\theta} \leftarrow -\nabla L_\theta(\hat{\theta}) \qquad (12)$$

To update parameter approximation $\hat{\theta}$.

$$\hat{\theta}_t \leftarrow \hat{\theta} + \hat{\theta}_t = \sum_{i=0}^{t} \hat{\theta}_i \qquad (13)$$

To save the final approximation $\hat{\theta}$.

$$\hat{\theta} = \sum_{i=0}^{M} \hat{\theta}_i \qquad (14)$$

Advantages of the method:
- The method is easy to implement.
- Iteratively corrects weak classifier errors and improves accuracy by combining vulnerable learners.
- Not prone to overtraining.

Disadvantages of the method:
- Sensitive to noisy data.
- The method is strongly affected by deviations in the data.

4.4. Models Accuracy Estimation Methods

To assess the adequacy of the models we used the relative error [80]. The relative error is the ratio of the absolute measurement error to the measurement performed.

$$RE = \frac{Absolute\ Error}{Measurement\ Being\ Taken} \qquad (15)$$

In the case of evaluating models on different samples with different values, the relative error allows us to estimate the accuracy in relative terms.

For use in public health practice models, the mean absolute error was calculated [81]. It is a measure of the error between the predicted and observed values.

$$MAE = \frac{\sum_{i=1}^{n} |y_i - x_i|}{n} \qquad (16)$$

where y_i is the predicted value, x_i is the observed value, n is the number of observations.

5. Results

Models of the COVID-19 epidemic process were implemented using the Python programming language. An experimental study of the models was carried out on data on new cases of COVID-19 presented in the Coronavirus Resource Center of Johns Hopkins University and Medicine for Germany, Japan, South Korea, and Ukraine. The forecast is built for 3, 7, 10, 14, 21, and 30 days.

5.1. Forecasting Results

The forecast results show the retrospective dynamics of new cases of COVID-19 in the selected area.

Figure 1 shows the results of predicting new cases of COVID-19 with a Random Forest model. Figure 2 shows the results of predicting new cases of COVID-19 with a K-Nearest Neighbors model. Figure 3 shows the results of predicting new cases of COVID-19 with a Gradient Boosting model. Figures 1–3 show the results of simulations for Germany, Japan, South Korea, and Ukraine.

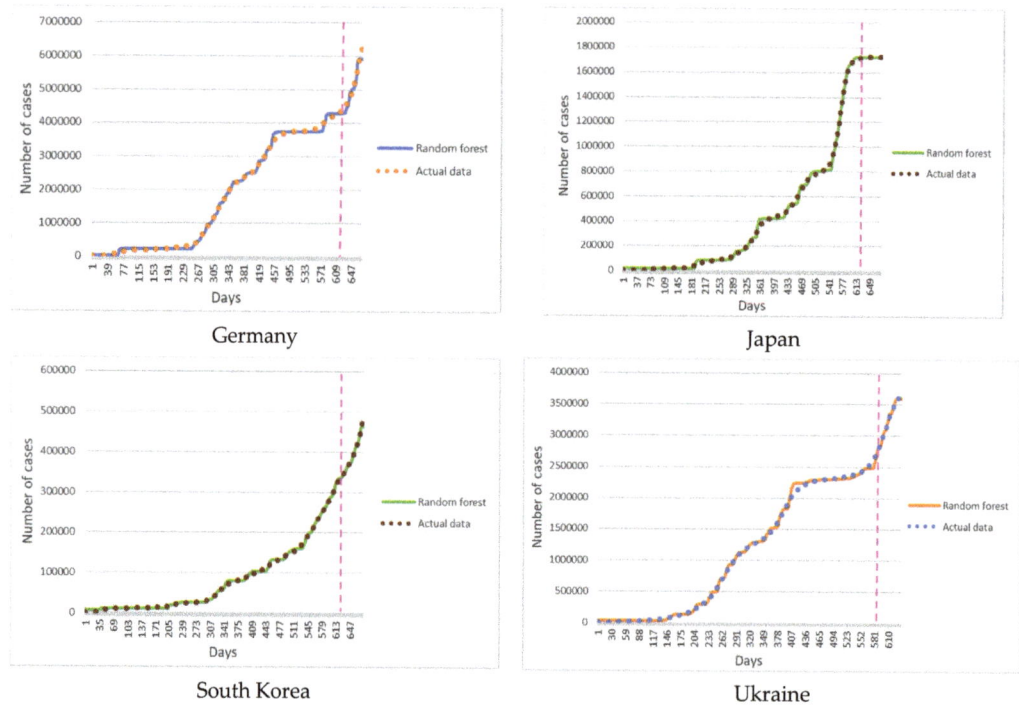

Figure 1. Forecasting of COVID-19 new cases by Random Forest model.

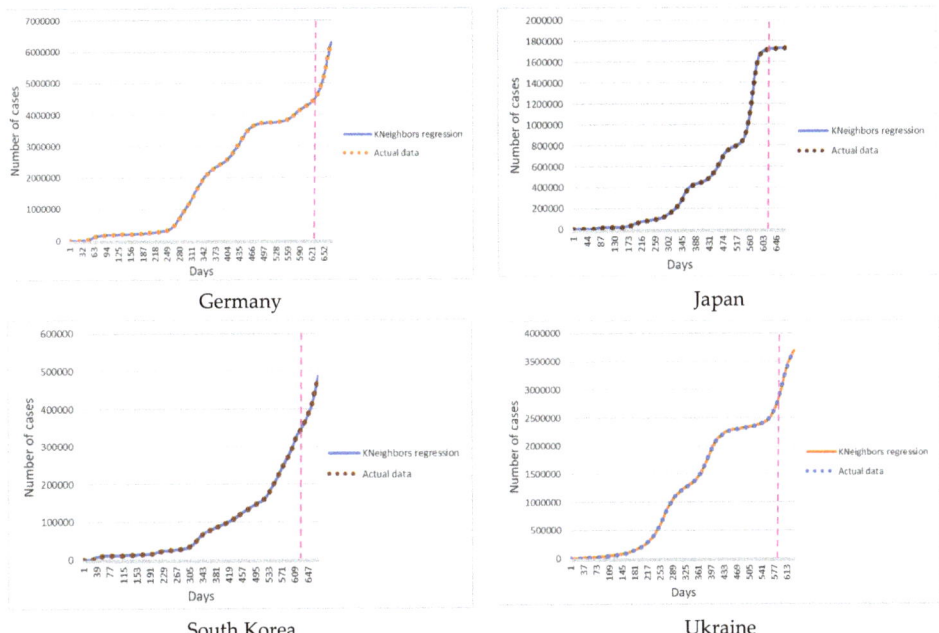

Figure 2. Forecasting of COVID-19 new cases by K-Nearest Neighbors model.

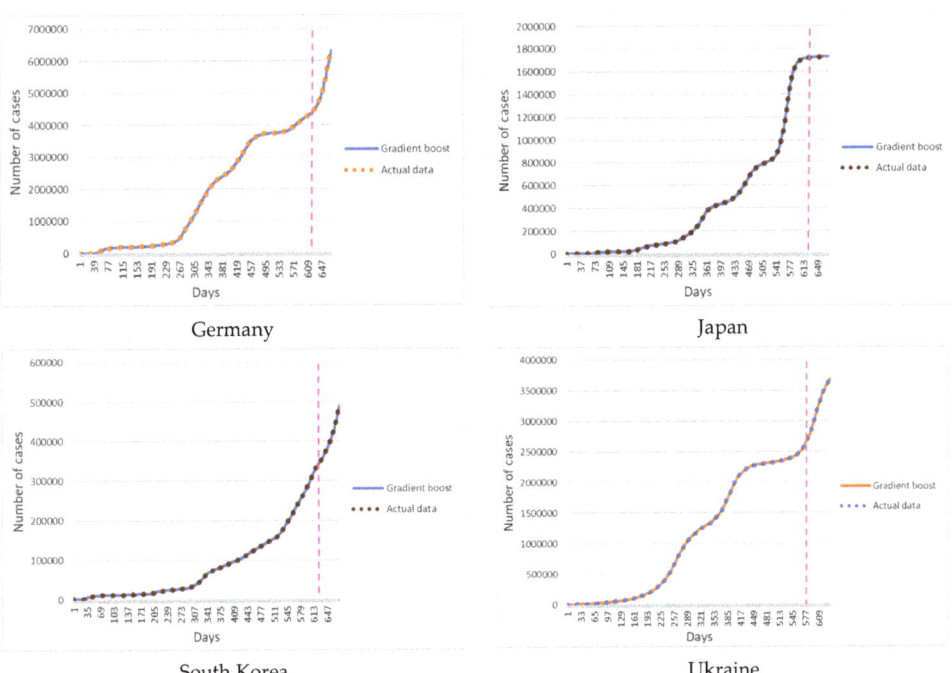

Figure 3. Forecasting of COVID-19 new cases by Gradient Boosting model.

5.2. Forecasting Accuracy Estimation

To assess the accuracy of the models, the relative error and the average absolute error were calculated for the retrospective forecast of the cumulative values of new cases of COVID-19 for the selected territories for 3, 7, 10, 14, 21, and 30 days. The relative error of training data shows the adequacy of the constructed model. The relative error of forecasted data shows the accuracy of the constructed model. However, the error in absolute incidence values is more informative for use in practice by epidemiologists and public health specialists. Absolute incidence rates make it possible to assess the future epidemic situation and take the necessary control measures to contain the epidemic.

Table 1 shows the relative error of developed models for predicting new cases of COVID-19 in Germany.

Table 1. Relative error of forecasted new cases for Germany (%).

Duration of Forecast (Days)	Random Forest Model	K-Nearest Neighbors Model	Gradient Boosting Model
Training 3	9.409708	0.927918	0.767574
Forecast 3	5.17024	0.544484	0.010204
Training 7	9.45157	0.930967	0.772348
Forecast 7	3.849152	0.473528	0.007183
Training 10	9.49197	0.932761	0.775975
Forecast 10	3.012969	0.491807	0.006026
Training 14	9.534235	0.934895	0.780697
Forecast 14	2.995393	0.517344	0.012878
Training 21	9.615585	0.93961	0.788661
Forecast 21	2.804224	0.510144	0.031779
Training 30	9.737732	0.947268	0.799813
Forecast 30	2.392481	0.474844	0.029858

Table 2 shows the relative error of developed models for predicting new cases of COVID-19 in Japan.

Table 2. Relative error of forecasted new cases for Japan (%).

Duration of Forecast (Days)	Random Forest Model	K-Nearest Neighbors Model	Gradient Boosting Model
Training 3	49.94565	0.858646	2.272102
Forecast 3	0.441742	0.003262	0.021831
Training 7	50.2621	0.863787	2.286198
Forecast 7	0.430174	0.002863	0.02099
Training 10	50.5022	0.867681	2.296893
Forecast 10	0.421757	0.002989	0.020306
Training 14	50.82606	0.872933	2.311317
Forecast 14	0.410848	0.002813	0.019649
Training 21	51.40349	0.882265	2.337009
Forecast 21	0.388179	0.003121	0.018877
Training 30	52.16673	0.894539	2.370913
Forecast 30	0.351485	0.003791	0.018152

Table 3 shows the relative error of developed models for predicting new cases of COVID-19 in South Korea.

Table 3. Relative error of forecasted new cases for South Korea (%).

Duration of Forecast (Days)	Random Forest Model	K-Nearest Neighbors Model	Gradient Boosting Model
Training 3	5.979816	0.944363	0.427686
Forecast 3	0.97913	0.374233	0.005938
Training 7	6.011627	0.947449	0.430314
Forecast 7	0.991853	0.406392	0.007046
Training 10	6.036437	0.949663	0.432075
Forecast 10	0.952094	0.421965	0.022028
Training 14	6.068961	0.953124	0.434636
Forecast 14	0.966735	0.409796	0.0236
Training 21	6.130652	0.9597	0.439033
Forecast 21	0.869629	0.386758	0.029737
Training 30	6.208069	0.969129	0.444362
Forecast 30	0.891858	0.356536	0.043103

Table 4 shows the relative error of developed models for predicting new cases of COVID-19 in Ukraine.

Table 4. Relative error of forecasted new cases for Ukraine (%).

Duration of Forecast (Days)	Random Forest Model	K-Nearest Neighbors Model	Gradient Boosting Model
Training 3	13.77781	1.313108	0.772439
Forecast 3	1.844497	0.118812	0.010837
Training 7	13.86759	1.32033	0.777511
Forecast 7	1.134243	0.154387	0.011266
Training 10	13.9369	1.326011	0.781295
Forecast 10	0.907004	0.149701	0.015239
Training 14	14.02436	1.333106	0.786291
Forecast 14	1.0065	0.171587	0.022496
Training 21	14.18892	1.345467	0.795291
Forecast 21	0.855615	0.197928	0.025963
Training 30	14.40387	1.361099	0.807296
Forecast 30	0.814586	0.22747	0.025859

Table 5 shows the mean absolute error of developed models for predicting cumulative new cases of COVID-19 in Germany.

Table 5. Mean absolute error of forecasted cumulative new cases for Germany (number of cases).

Duration of Forecast (Days)	Random Forest Model	K-Nearest Neighbors Model	Gradient Boosting Model
Forecast 3	323,198.7	34,082.67	639
Forecast 7	238,495.9	29,184.86	445.4286
Forecast 10	185,499.4	29,834.6	370.4
Forecast 14	180,502	107,511	757.2857
Forecast 21	163,443	29,349.67	1748.524
Forecast 30	135,644.6	26,417.57	1598.367

Table 6 shows the mean absolute error of developed models for predicting cumulative new cases of COVID-19 in Japan.

Table 6. Mean absolute error of forecasted cumulative new cases for Japan (number of cases).

Duration of Forecast (Days)	Random Forest Model	K-Nearest Neighbors Model	Gradient Boosting Model
Forecast 3	7628.333	56.33333	377
Forecast 7	7427.714	49.42857	362.4286
Forecast 10	7281.8	51.6	350.6
Forecast 14	7092.714	52.8765	339.2143
Forecast 21	6700	53.85714	325.8095
Forecast 30	6064.967	65.4	313.2

Table 7 shows the mean absolute error of developed models for predicting cumulative new cases of COVID-19 in South Korea.

Table 7. Mean absolute error of forecasted cumulative new cases for South Korea (number of cases).

Duration of Forecast (Days)	Random Forest Model	K-Nearest Neighbors Model	Gradient Boosting Model
Forecast 3	4518	1731.333	27.33333
Forecast 7	4495.286	1840.714	32
Forecast 10	4258.2	1883.8	96.4
Forecast 14	4249.357	3081.953	101.7143
Forecast 21	3737.238	1659.571	123.7619
Forecast 30	3698.833	1491.267	172.9

Table 8 shows the mean absolute error of developed models for predicting cumulative new cases of COVID-19 in Ukraine.

Table 8. Mean absolute error of forecasted cumulative new cases for Ukraine (number of cases).

Duration of Forecast (Days)	Random Forest Model	K-Nearest Neighbors Model	Gradient Boosting Model
Forecast 3	67,517.67	4343.333	396.6667
Forecast 7	41,389.43	5601.714	409.1429
Forecast 10	33,009.6	5409.4	549.8
Forecast 14	31,624.25	6201.8723	756.23
Forecast 21	30,471.67	6962.619	913
Forecast 30	28,407.57	7791.3	886.5333

5.3. Models Complexity Estimation

Let us estimate the computational complexity of the Random Forest model. When building a model, it has a large size. The complexity of the model is $O(NK)$, where N is the number of trees.

The complexity of training the K-Nearest Neighbors model is $O(1)$. $O(n)$ is technically correct as well. It is needed to remember the training sample. Prediction complexity is $O(n)$ for each feature. If it is required to predict k objects independently using a fixed training sample, then the complexity will be $O(kn)$.

The complexity of the Gradient Boosting model is $O(M n \log_n d)$, where M is the number of trees. In general, the model takes longer than a Random Forest because it builds the next tree based on the error or residual of the previous tree, so the process cannot be parallelized compared to a Random Forest.

6. Discussion

It should be noted that COVID-19 refers to infections with an easily possible aerosol transmission mechanism of the pathogen, the source of which is a sick person and a carrier, i.e., an asymptomatic person who sheds a pathogen into the environment and infects other susceptible people. The epidemic process of such infections is significantly influenced by social factors, such as crowding, physical distancing, mask regimen, vaccination coverage of the population, etc. [82]. A step-by-step assessment of the predicted morbidity and its comparison with the registered one allows not only to correctly assess the epidemic situation, the manifestations of the epidemic process characteristic of specific conditions of space and time, but also to assess the quality, effectiveness, and correctness of the preventive and anti-epidemic measures taken, to choose the optimal ones on time and make adjustments as in regulatory documents, and in local preventive action plans.

New challenges for humanity associated with the COVID-19 pandemic forced specialists from various fields of science to mobilize their capabilities. The contribution of specialists in mathematical modeling can be essential for studying the dynamics and characteristics of the manifestations of the epidemic process of emergent infection, the behavior of the pathogen, and the patterns of the spread of the disease are studied simultaneously with the development of preventive and anti-epidemic measures [83]. For a clearer understanding of the patterns of the spread of the COVID-19 pathogen and the choice of the most meaningful and rational measures, we propose evaluating the forecast results through different periods. This information will make it possible to understand the dynamics and features of the epidemic process characteristic of a specific time and a specific territory for which the forecast is made.

The first step is to estimate the expected incidence of COVID-19 after 3 days. The results obtained do not yet allow assessing the correctness of management decisions and the effectiveness of the measures that have been implemented. However, we can understand whether the intensity of the epidemic process has changed compared to the period for which case data were used to build a forecast. Lower rates of predicted morbidity than the actual ones indicate the intensification of the epidemic process and the need to strengthen control measures, which should be paid attention to by decision-makers. The disadvantage of this forecast is that if a period is taken that includes weekends and holidays, then the excess of the predicted incidence compared to the registered one will not reflect the effectiveness of the measures taken. The actual incidence may significantly exceed the registered one [84].

The second step may be to assess the incidence after 7 days. The forecast results after this period allow us to give a preliminary assessment of the correctness of the adopted management decisions. Considering that the average incubation period of COVID-19 is 5–6 days [85], the excess of the actual incidence data of the predicted incidence indicators will roughly give an idea of the need to strengthen control measures, draw the attention of decision-makers to the quality and correctness of the measures that have been developed. An approximate judgment can also be made about the amount of medical care needed for the population. The forecast after 7 days also allows to smooth out the error associated with holidays.

The third step compares the predicted and actual morbidity after 10 days, making it possible to assess the correctness of management decisions more accurately [86]. Fluctuations in incidence associated with weekends and holidays will be leveled. Cases in which infection occurred when the modeling was carried out will be registered. The driving forces of the epidemic process that were in effect for that period (cases with an average incubation period) were taken into account, so those cases that arose after the time when the model was built. New factors could arise or become more active that affect the dynamics and intensity of the COVID-19 epidemic process.

The next step is to assess the incidence in two weeks. 14 days is the maximum incubation period [87]. All cases of infection that occurred at the time of forecasting will already manifest as morbidity or carriage. Comparison of predicted and registered morbidity will allow assessing changes in the dynamics and intensity of the epidemic process, assessing the quality and effectiveness of the measures taken and the correctness of the managerial decisions made, and, if necessary, making adjustments to the volume and content of the control and preventive measures taken. In addition, in two weeks, it is possible to adjust the medical and laboratory network [88]. Exceeding the predicted indicators after 14 days of those indicators registered on the modeling day is a signal for drawing up plans to deploy additional beds for patients, including beds equipped with oxygen, purchase the necessary diagnostic test systems, medicines, and train medical personnel. It is also a signal to strengthen the vaccination campaign in the territory [89].

The next step is to evaluate the forecast data after 21 days. The results allow us to assess the epidemic situation and be a warning for time-taking measures to correct the situation, if necessary. Increasing rates of morbidity growth are a marker for the development of additional measures. You can also preliminarily estimate the required amount of resources—test systems for diagnostics, beds, oxygen stations, medicines, and medical personnel and understand whether the activities included in the plan at the previous stage were sufficient.

Furthermore, finally, the sixth step can be to assess the forecast of incidence in 30 days, which first of all, allows us to assess the burden on the healthcare system, institutions that provide medical care, the required amount of resources and personnel, and the damage from this disease [90]. Estimating the predicted morbidity within this period allows for the taking of necessary advance measures to manage peaks or extreme indicators, such as providing institutions with the necessary resources, and conducting training and retraining of medical personnel, considering the current situation. Other possible strategies include developing the optimal logistics for medical support of both patients and healthy individuals to be vaccinated (organization of vaccination points, providing training of vaccination teams, development of routes, purchase of vaccines, etc.).

To choose a simulation method, one should also consider the possibility of retraining machine learning models. Retraining is characterized by a significant excess of the error value of the test sample of the value of the average error of the training sample. An analysis of the models built in the framework of this study showed that all models are not overfitted.

The minimum number of observations required for a correct result was also analyzed. For a model based on the Random Forest method, the minimum required number of observations is 40, for the Gradient Boosting model—25, for the K-Nearest Neighbors model—15.

7. Conclusions

The paper describes the results of experimental studies of three models based on statistical machine learning methods: Random Forest, K-Nearest Neighbors, and Gradient Boosting. The experiments were performed on new COVID-19 case data provided by the Coronavirus Resource Center of Johns Hopkins University and Medicine for Germany, Japan, South Korea, and Ukraine. These countries were selected because they have different dynamics of the epidemic process and different measures that health systems have implemented to control the pandemic.

All models showed sufficient accuracy in deciding to implement control measures to counter the COVID-19 pandemic. The tasks that can be solved with the help of models depending on the period of the constructed predictive incidence are described.

The prediction accuracy of the Random Forest model is from 94.83% to 99.65%, the K-Nearest Neighbors models are from 99.46% to 99.96%, and the Gradient Boosting models are from 99.97% to 99.99%.

An analysis of the change in the error depending on the forecasting period showed a high agreement between the registered and actual statistics on the incidence of COVID-19 in Japan and South Korea, a satisfactory agreement between the data in Germany, and a low agreement between the registered and actual incidence of COVID-19 in Ukraine. This is due to the completeness of population testing and the testing approaches those countries have implemented during the pandemic.

The scientific novelty of the study lies in the development and study of models of emerging infections using the example of COVID-19 based on simple methods of statistical machine learning. In contrast to other studies, the article analyzes various periods for constructing a forecast, which makes it possible to evaluate the effectiveness of its use for solving various problems of public health.

The practical novelty of the study lies in the implementation of an automated tool for assessing the dynamics of the COVID-19 epidemic process in various territories. It is shown what tasks of epidemiology can be solved when building forecasts for various periods. The accuracy of modeling depends on the completeness of the data of the recorded statistics. Another essential practical value is the ability of public health experts to make decisions based only on new cases of COVID-19. This is especially true for areas where collecting other patient data is not possible due to low funding for the healthcare system or force majeure. For example, in Russia's war in Ukraine, it is impossible to collect complete data on COVID-19 cases, especially in the temporarily occupied territories and territories where active hostilities are taking place. Under such conditions, the proposed approach will be practical for the timely control of the COVID-19 epidemic process.

Future research development. Despite the high accuracy of the epidemic process models developed in the framework of this study based on statistical machine learning methods, such models do not allow us to identify the factors that affect the development of the epidemic process. It is the identification of factors and assessing their informativity that is an essential task of public health. Therefore, a further development of the study would combine the proposed machine learning models with multi-agent models of epidemic processes. On the one hand, multi-agent models will make it possible to identify and evaluate the factors influencing the dynamics of the epidemic process. On the other hand, machine learning models will improve the accuracy of the predictive incidence of multi-agent models. This will improve the adequacy of experimental studies and the effectiveness of decisions made based on simulation.

Author Contributions: Conceptualization, D.C. and I.M.; methodology, D.C.; software, I.M. and K.B.; validation, D.C., I.M. and T.C.; formal analysis, D.C. and T.C.; investigation, D.C. and S.Y.; resources, D.C., I.M. and T.C.; writing—original draft preparation, D.C.; writing—review and editing, I.M., K.B., T.C. and S.Y.; visualization, I.M.; supervision, D.C.; project administration, S.Y. All authors have read and agreed to the published version of the manuscript.

Funding: The study was funded by the National Research Foundation of Ukraine in the framework of the research project 2020.02/0404 on the topic "Development of intelligent technologies for assessing the epidemic situation to support decision-making within the population biosafety management".

Institutional Review Board Statement: Not applicable.

Informed Consent Statement: Not applicable.

Data Availability Statement: The initial data used in this research is publicly available in Coronavirus Resource Center of Johns Hopkins University and Medicine by link https://coronavirus.jhu.edu/ (accessed on 25 April 2022).

Conflicts of Interest: The authors declare no conflict of interest.

References

1. Carvalho, T.; Krammer, F.; Iwasaki, A. The first 12 months of COVID-19: A timeline of immunological insights. *Nat. Rev. Immunol.* **2021**, *21*, 245–256. [CrossRef] [PubMed]
2. Liu, Q.; Xu, K.; Wang, X.; Wang, W. From SARS to COVID-19: What lessons have we learned? *J. Infect. Public Health* **2020**, *13*, 1611–1618. [CrossRef] [PubMed]
3. Khan, M.; Adil, S.F.; Alkhathlan, H.Z.; Tahir, M.N.; Saif, S.; Khan, M.; Khan, S.T. COVID-19: A global challenge with old history, epidemiology and progress so far. *Molecules* **2020**, *26*, 39. [CrossRef] [PubMed]
4. Shih, H.I.; Wu, C.J.; Tu, Y.F.; Chi, C.Y. Fighting COVID-19: A quick review of diagnoses, therapies, and vaccines. *Biomed. J.* **2020**, *43*, 341–354. [CrossRef]
5. Chen, X.; Gong, W.; Wu, X.; Zhao, W. Estimating economic losses caused by COVID-19 under multiple control measure scenarios with a coupled infectious disease-economic model: A case study in Wuhan, China. *Int. J. Environ. Res. Public Health* **2021**, *18*, 11753. [CrossRef]

6. Branquinho, C.; Santos, A.C.; Noronha, C.; Ramiro, L.; de Matos, M.G. COVID-19 pandemic and the second lockdown: The 3rd wave of the disease through the voice of youth. *Child Indic. Res.* **2022**, *15*, 199–216. [CrossRef]
7. Hossain, K.; Hassanzadeganroudsari, M.; Apostolopoulos, V. The emergence of new strains of SARS-CoV-2. What does it mean for COVID-19 vaccines? *Expert Rev. Vaccines* **2021**, *20*, 635–638. [CrossRef]
8. Brüssow, H. COVID-19: Vaccination problems. *Environ. Microbiol.* **2021**, *23*, 2878–2890. [CrossRef]
9. Begum, J.; Mir, N.A.; Dev, K.; Buyamayum, B.; Wani, M.Y.; Raza, M. Challenges and prospects of COVID-19 vaccine development based on the progress made in SARS and MERS vaccine development. *Transbound. Emerg. Dis.* **2021**, *68*, 1111–1124. [CrossRef]
10. Shan, S.G.S.; Nogueras, D.; van Woerden, H.C.; Kiparoglou, V. The COVID-19 pandemic: A pandemic of lockdown loneliness and the role of digital technology. *J. Med. Internet Res.* **2020**, *22*, e22287. [CrossRef]
11. Fedorovich, O.; Uruskiy, O.; Pronchakov, Y.; Lukhanin, M. Method and information technology to research the component architecture of products to justify investments of high-tech enterprise. *Radioelectron. Comput. Syst.* **2021**, *1*, 150–157. [CrossRef]
12. Agosto, A.; Giudici, P. COVID-19 contagion and digital finance. *Digit. Financ.* **2020**, *2*, 159–167. [CrossRef] [PubMed]
13. Fedushko, S.; Ustyianovych, T. E-commerce customers behavior research using cohort analysis: A case study of COVID-19. *J. Open Innov. Technol. Mark. Complex.* **2022**, *8*, 12. [CrossRef]
14. Davidich, N.; Chumachenko, I.; Davidich, Y.; Hanieva, T.; Artsybasheva, N.; Melenchuk, T. Advanced traveller information systems to optimizing freight driver route selection. In Proceedings of the 2020 13th International Conference on Developments in eSystems Engineering (DeSE), Liverpool, UK, 14–17 December 2020; pp. 111–115. [CrossRef]
15. Misiuk, T.; Kondratenko, Y.; Sidenko, I.; Kondratenko, G. Computer vision mobile system for education using augmented reality technology. *J. Mob. Multimed.* **2021**, *17*, 555–576.
16. Nechyporenko, A.; Reshetnik, V.; Shyian, D.; Yurevych, N.; Alekseeva, V.; Nazaryan, R.; Gargin, V. Comparative characteristics of the anatomical structures of the ostiomeatal complex obtained by 3D modeling. In Proceedings of the 2020 IEEE International Conference on Problems of Infocommunications Science and Technology, Kharkiv, Ukraine, 6–9 October 2021; pp. 407–411. [CrossRef]
17. Bazilevych, K.; Krivtsov, S.; Butkevych, M. Intelligent evaluation of the informative features of cardiac studies diagnostic data using Shannon method. *CEUR Workshop Proc.* **2021**, *3003*, 65–75.
18. Izonin, I.; Tkachenko, R.; Verhun, V.; Zub, K. An approach towards missing data management using improved GRNN-SGTM ensemble method. *Eng. Sci. Technol.* **2021**, *24*, 749–759. [CrossRef]
19. Yakovlev, S.; Bazilevych, K.; Chumachenko, D.; Chumachenko, T.; Hulianytskyi, L.; Meniailov, I.; Tkachenko, A. The concept of developing a decision support system for the epidemic morbidity control. *CEUR Workshop Proc.* **2020**, *2753*, 265–274.
20. Ross, R. An application of the theory of probabilities to the study of a priori pathometry. *Proc. R. Soc. Lond.* **1916**, *92*, 204–230. [CrossRef]
21. Hamer, W. The Milroy lectures. On epidemic disease in England—The evidence of variability and of persistency of type. *Lancet* **1906**, *167*, 569–574. [CrossRef]
22. Kermack, W.O.; McKendrick, A.G. Contribution to the mathematical theory to epidemics. *Proc. R. Soc. Lond.* **1927**, *115*, 700–721. [CrossRef]
23. Holz, M.; Fahr, A. Compartment modeling. *Adv. Drug Deliv. Rev.* **2001**, *48*, 249–264. [CrossRef]
24. Kuddus, A.; Mohiuddin, M.; Rahman, A. Mathematical analysis of a measles transmission dynamics model in Bangladesh with double dose vaccination. *Sci. Rep.* **2021**, *11*, 16571. [CrossRef] [PubMed]
25. Ostus, D.; Hickmann, K.S.; Caragea, P.C.; Higdon, D.; del Valle, S.Y. Forecasting seasonal influenza with a state-space SIR model. *Ann. Appl. Stat.* **2017**, *11*, 202–224. [CrossRef] [PubMed]
26. Baroyan, O.V.; Rvachev, L.A. Deterministic models of epidemics for a territory with a transport network. *Cybern. Syst. Snalysis* **1967**, *3*, 55–61. [CrossRef]
27. Rihan, F.A.; Baleanu, D.; Lakshmanan, S.; Rakkiyappan, R. On fractional SIRC model with Salmonella bacterial infection. *Abstr. Appl. Anal.* **2014**, *2014*, 136263. [CrossRef]
28. Van Effelterre, T.P.; Zink, T.K.; Hoet, B.J.; Hausdorff, W.P.; Rosenthal, P. A mathematical model of Hepatitis A transmission in the United States indicates value of universal childhood immunization. *Clin. Infect. Dis.* **2006**, *43*, 158–164. [CrossRef]
29. Islam, Z.; Ahmed, S.; Rahman, M.M.; Karim, M.F.; Amin, M.R. Global stability and parameter estimation for a diphtheria model: A case study of an epidemic in Rohingya refugee camp in Bangladesh. *Comput. Math. Methods Med.* **2022**, *2022*, 6545179. [CrossRef]
30. Huo, H.F.; Chen, R.; Wang, X.Y. Modelling and stability of HIV/AIDS epidemic model with treatment. *Appl. Math. Model.* **2016**, *40*, 6550–6559. [CrossRef]
31. Khan, M.A.; Islam, S.; Arif, M.; ul Haq, Z. Transmission model of Hepatitis B virus with the migration effect. *BioMed Res. Int.* **2013**, *2013*, 150681. [CrossRef]
32. Stocks, T.; Martin, L.J.; Kuhlmann-Berenzon, S.; Britton, T. Dynamic modeling of Hepatitis C transmission among people who inject drugs. *Epidemics* **2020**, *30*, 100378. [CrossRef] [PubMed]
33. Brown, R.A. A simple model for control of COVID-19 infections on an urban campus. *Proc. Natl. Acad. Sci. USA* **2021**, *118*, e2105292118. [CrossRef] [PubMed]
34. Zha, J.; Liu, X.; Sui, M. The SEIR model of COVID-19 forecasting rates of infections in New York city. *J. Phys. Conf. Ser.* **2021**, *1735*, 012009. [CrossRef]

35. Radulescu, A.; Williams, C.; Cavanagh, K. Management strategies in a SEIR-type model of COVID-19 community spread. *Sci. Rep.* **2020**, *10*, 21256. [CrossRef] [PubMed]
36. Erandi, K.K.W.H.; Mahasinghe, A.C.; Perera, S.S.N.; Jayasinghe, S. Effectiveness of the strategies implemented in Sri Lanka for controlling the COVID-19 outbreak. *J. Appl. Math.* **2020**, *2020*, 2954519. [CrossRef]
37. Alexandrou, C.; Harmandaris, V.; Irakleous, A.; Koutsou, G.; Savva, N. Modeling the evolution of COVID-19 via compartmental and particle-based approaches: Application to the Cyprus case. *PLoS ONE* **2020**, *16*, e0250709. [CrossRef] [PubMed]
38. Forien, R.; Pang, G.; Pardoux, E. Estimating the state of the COVID-19 epidemic in France using a model with memory. *R. Soc. Open Sci.* **2021**, *8*, 202327. [CrossRef]
39. Nistal, R.; de la Sen, M.; Gabirondo, J.; Alonso-Quesada, S.; Garrido, A.J.; Garrido, I. A Study on COVID-19 Incidence in Europe through Two SEIR Epidemic Models Which Consider Mixed Contagions from Asymptomatic and Symptomatic Individuals. *Appl. Sci.* **2021**, *11*, 6266. [CrossRef]
40. Cooped, I.; Mondal, A.; Antonopoulos, C.G. A SIR model assumption for the spread of COVID-19 in difference communities. *Chaos Solut. Fractals* **2020**, *139*, 110057. [CrossRef]
41. Al-Raeei, M.; El-Daher, M.S.; Solieva, O. Applying SEIR model without vaccination for COVID-19 in case of the United States, Russia, the United Kingdom, Brazil, France, and India. *Epidemiol. Methods* **2021**, *10*, 20200036. [CrossRef]
42. Odagaki, T. Exact properties of SIQR model for COVID-19. *Phys. A* **2021**, *564*, 125564. [CrossRef]
43. Mugdha, S.B.S.; Uddin, M.; Islam, T. Extended epidemiological models for weak economic region: Case studies of the spreading of COVID-19 in the South Asian subcontinental countries. *BioMed Res. Int.* **2021**, *2021*, 7787624. [CrossRef] [PubMed]
44. Rahimi, I.; Gandomi, A.H.; Asteris, P.G.; Chen, F. Analysis and Prediction of COVID-19 Using SIR, SEIQR, and Machine Learning Models: Australia, Italy, and UK Cases. *Information* **2021**, *12*, 109. [CrossRef]
45. Franco, N. COVID-19 Belgium: Extended SEIR-QD model with nursing homes and long-term scenarios-based forecasts. *Epidemics* **2021**, *37*, 100490. [CrossRef] [PubMed]
46. Ghostine, R.; Gharamti, M.; Hassrouny, S.; Hoteit, I. An Extended SEIR Model with Vaccination for Forecasting the COVID-19 Pandemic in Saudi Arabia Using an Ensemble Kalman Filter. *Mathematics* **2021**, *9*, 636. [CrossRef]
47. Gorbachuk, V.M.; Dunaievskyi, M.S.; Syrku, A.A.; Suleimanov, S.B. Substantiating the diffusion model of innovation implementation and its application to vaccine propagation. *Cybern. Syst. Anal.* **2022**, *58*, 84–94. [CrossRef]
48. Guan, L.; Prieur, C.; Zhang, L.; Prieur, C.; Georges, D.; Bellemain, P. Transport effect of COVID-19 pandemic in France. *Annu. Rev. Control* **2020**, *50*, 394–408. [CrossRef]
49. Putra, Z.A.; Abidin, S.A.Z. Application of SEIR model in COVID-19 and the effect of lockdown on reducing the number of active cases. *Indones. J. Sci. Technol.* **2020**, *5*, 10–17. [CrossRef]
50. Dashtbali, M.; Mirzaie, M. A compartmental model that predicts the effect of social distancing and vaccination on controlling COVID-19. *Sci. Rep.* **2021**, *11*, 8191. [CrossRef]
51. Morciglio, A.; Zhang, B.; Chowell, G.; Hyman, J.M.; Jiang, Y. Mask-Ematics: Modeling the Effects of Masks in COVID-19 Transmission in High-Risk Environments. *Epidemiologia* **2021**, *2*, 207–226. [CrossRef]
52. Asamoah, J.K.K.; Jin, Z.; Sun, G.Q.; Seidu, B.; Yankson, E.; Abidemi, A.; Oduro, F.T.; Moore, S.E.; Okyere, E. Sensitivity assessment and optimal economic evaluation of a new COVID-19 compartmental epidemic model with control interventions. *Chaos Solut. Fractals* **2021**, *146*, 110885. [CrossRef]
53. Masandawa, L.; Mirau, S.S.; Mbalawata, I.S. Mathematical modeling of COVID-19 transmission dynamics between healthcare workers and community. *Results Phys.* **2021**, *29*, 104731. [CrossRef] [PubMed]
54. Jen, G.H.H.; Chen, S.Y.; Chang, W.J.; Chen, C.N.; Yen, A.M.F.; Chang, R.E. Evaluating medical capacity for hospitalization and intensive care unit of COVID-19: A queue model approach. *J. Formos. Med. Assoc.* **2021**, *120* (Suppl. S1), S86–S94. [CrossRef]
55. Dong, E.; Du, H.; Gardner, L. An interactive web-based dashboard to track COVID-19 in real time. *Lancet Infect. Dis.* **2020**, *20*, 533–534. [CrossRef]
56. Robert Koch-Institut: COVID-19 Dashboard. 2020. Available online: https://experience.arcgis.com/experience/478220a4c45448 0e823b17327b2bf1d4 (accessed on 25 April 2022).
57. Schilling, J.; Tolksdorf, K.; Marquis, A.; Faber, M.; Pfoch, T.; Buda, S.; Haas, W.; Schuler, E.; Altmann, D.; Grote, U.; et al. Die verschiedenen Phasen der COVID-19-Pandemie in Deutschland: Eine deskriptive Analyse von Januar 2020 bis Februar 2021. *Bundesgesundheitsblat—Gesundh. Gesundh.* **2021**, *64*, 1093–1106. [CrossRef] [PubMed]
58. Braun, P. Effect of lockdown and vaccination on the course of the COVID-19 pandemic in Germany. *Int. J. Clin. Pharmacol. Ther.* **2022**, *60*, 125–135. [CrossRef]
59. Bittmann, S. Role of Omicron variant of SARS-CoV-2 in children in Germany. *World J. Pediatrics* **2022**, *18*, 283–284. [CrossRef]
60. Gandjour, A. How Many Intensive Care Beds are Justifiable for Hospital Pandemic Preparedness? A Cost-effectiveness Analysis for COVID-19 in Germany. *Appl. Health Econ. Health Policy* **2021**, *19*, 181–190. [CrossRef]
61. Leithauser, N.; Schneider, J.; Johann, S.; Krumke, S.O.; Schmidt, E.; Streicher, M.; Scholz, S. Quantifying Covid19-vaccine location strategies for Germany. *BMC Health Serv. Res.* **2021**, *21*, 780. [CrossRef]
62. Hara, Y.; Yamaguchi, H. Japanese travel behavior trends and change under COVID-19 state-of-emergency declaration: Nationwide observation by mobile phone location data. *Transp. Res. Interdiscip. Perspect.* **2021**, *9*, 100288. [CrossRef]
63. Ishikawa, Y.; Hifumi, T.; Urashima, M. Critical care medical centers may play an important role in reducing the risk of COVID-19 death in Japan. *SN Compr. Clin. Med.* **2020**, *2*, 2147–2150. [CrossRef]

64. Wu, Y.C.; Chen, C.S.; Chan, Y.J. The outbreak of COVID-19: An overview. *J. Chin. Med. Assoc.* **2020**, *83*, 217–220. [CrossRef] [PubMed]
65. Liu, S.; Yamamoto, T. Role of stay-at-home requests and travel restrictions in preventing the spread of COVID-19 in Japan. *Transp. Res. Part A Policy Pract.* **2022**, *159*, 1–16. [CrossRef] [PubMed]
66. Machida, M.; Nakamura, I.; Kojima, T.; Saito, R.; Nakaya, T.; Hanibuchi, T.; Takamiya, T.; Odagiri, Y.; Fukushima, N.; Kikuchi, H.; et al. Acceptance of a COVID-19 vaccine in Japan during the COVID-19 pandemic. *Vaccines* **2021**, *9*, 210. [CrossRef] [PubMed]
67. Kim, H. COVID-19 Apps as a digital intervention police: A longitudinal panel data analysis in South Korea. *Health Policy* **2021**, *125*, 1430–1440. [CrossRef]
68. Kim, N.; Kang, S.J.; Tak, S. Reconstructing a COVID-19 outbreak within a religious group using social network analysis simulation in Korea. *Epidemiol. Health* **2021**, *43*, e2021068. [CrossRef]
69. Lee, D.; Heo, K.; Seo, Y.; Ahn, H.; Jung, K.; Lee, S.; Choi, H. Flattening the curve on COVID-19: South Korea's measures in tackling initial outbreak of Coronavirus. *Am. J. Epidemiol.* **2021**, *190*, 496–505. [CrossRef]
70. Kim, Y.J.; Sung, H.; Ki, C.S.; Hur, M. COVID-19 testing in South Korea: Current status and the need for faster diagnostics. *Ann. Lab. Med.* **2020**, *40*, 349–350. [CrossRef]
71. Choi, Y.; Kim, J.S.; Kim, J.E.; Choi, H.; Lee, C.H. Vaccination prioritization strategies for COVID-19 in Korea: A mathematical modeling approach. *Int. J. Environ. Res. Public Health* **2021**, *18*, 4240. [CrossRef]
72. Gankin, Y.; Nemira, A.; Koniukhovskii, V.; Chowell, G.; Weppelmann, T.A.; Skums, P.; Kirpich, A. Investigating the first stage of the COVID-19 pandemic in Ukraine using epidemiological and genomic data. *Infect. Genet. Evol. J. Mol. Epidemiol. Evol. Genet. Infect. Dis.* **2021**, *95*, 105087. [CrossRef]
73. Ivats-Chabina, A.R.; Korolchuk, O.L.; Kachur, A.Y.; Smiianov, V.A. Healthcare in Ukraine during the pandemic: Difficulties, challenges and solutions. *Wiad. Lek.* **2021**, *74*, 1256–1261. [CrossRef]
74. Patel, S.S.; Moncayo, O.E.; Conroy, K.M.; Jordan, D.; Erickson, T.B. The landscape of disinformation of health crisis communication during the COVID-19 pandemic in Ukraine: Hybrid warfare tactics, fake media news and review of evidence. *J. Sci. Commun.* **2020**, *19*, AO2. [CrossRef] [PubMed]
75. Matiashova, L.; Isayeva, G.; Shanker, A.; Tsagkaris, C.; Aborode, A.T.; Essar, M.Y.; Ahmad, S. COVID-19 vaccination in Ukraine: An update on the status of vaccination and the challenges at hand. *J. Med. Virol.* **2021**, *93*, 5252–5253. [CrossRef] [PubMed]
76. Guerard, J.B. Regression analysis and forecasting models. In *Introduction to Financial Forecasting in Investment Analysis*; Springer: New York, NY, USA, 2013. [CrossRef]
77. Svetnik, V.; Liaw, A.; Tong, C.; Culberson, J.C.; Sheridan, R.P.; Feuston, B.P. Random Forest: A classification and regression tool for compound classification and QSAR modeling. *J. Chem. Inf. Comput. Sci.* **2003**, *43*, 1947–1958. [CrossRef] [PubMed]
78. Kumar, T. Solution of linear and non linear regression problem by K Nearest Neighbour approach: By using three sigma rule. In Proceedings of the 2015 IEEE International Conference on Computational Intelligence & Communication Technology, Ghaziabad, India, 13–14 February 2015; pp. 197–201. [CrossRef]
79. Singh, U.; Rizwan, M.; Alaraj, M.; Alsaidan, I. A machine learning-based gradient boosting regression approach for wind power production forecasting: A step towards smart grid environments. *Energies* **2021**, *14*, 5196. [CrossRef]
80. Chen, C.; Twycross, J.; Garibaldi, J.M. A new accuracy measure based on bounded relative error for time series forecasting. *PLoS ONE* **2017**, *12*, e0174202. [CrossRef]
81. Willmott, C.J.; Matsuura, K. Advantages of the mean absolute error (MAE) over the root mean square error (RMSE) in assessing average model performance. *Clim. Res.* **2005**, *30*, 79–82. [CrossRef]
82. Seligman, B.; Ferranna, M.; Bloom, D.E. Social determinants of mortality from COVID-19: A simulation study using NHANES. *PLoS Med.* **2021**, *18*, e1003490. [CrossRef]
83. Rubin, D.M.; Achari, S.; Carlson, C.S.; Letts, R.F.R.; Pantanowitz, A.; Postema, M.; Richards, X.L.; Wigdorowitz, B. Facilitating understanding, modeling and simulation of infectious disease epidemics in the age of COVID-19. *Front. Public Health* **2021**, *9*, 593417. [CrossRef]
84. Wiwanitkit, V.; Joob, B. Density of COVID-19 and mass population movement during long holiday: Simulation comparing between using holiday postponement and no holiday postponement. *J. Res. Med. Sci. Off. J. Isfahan Univ. Med. Sci.* **2020**, *25*, 55. [CrossRef]
85. Lauer, A.S.; Grantz, K.H.; Bi, Q.; Jones, F.K.; Zheng, Q.; Meredith, H.R.; Azman, A.S.; Reich, N.G.; Lessler, J. The incubation period of coronavirus disease 2019 (COVID-19) from publicly reported confirmed cases: Estimation and application. *Ann. Intern. Med.* **2020**, *172*, 577–582. [CrossRef]
86. Chen, Y.; Li, Q.; Karimian, H.; Chen, X.; Li, X. Spatio-temporal distribution characteristics and influencing factors of COVID-19 in China. *Sci. Rep.* **2021**, *11*, 3717. [CrossRef] [PubMed]
87. Leung, C. The incubation period of COVID-19: Current understanding and modeling technique. *Adv. Exp. Med. Biol.* **2021**, *1318*, 81–90. [CrossRef] [PubMed]
88. Zeinalnezhad, M.; Chofreh, A.G.; Goni, F.A.; Klemes, J.J.; Sari, E. Simulation and improvement of patients' workflow in heart clinics during COVID-19 pandemic using timed coloured Petri nets. *Int. J. Environ. Res. Public Health* **2020**, *17*, 8577. [CrossRef] [PubMed]

89. Karabay, A.; Kuzdeuov, A.; Varol, H.A. COVID-19 vaccination strategies considering hesitancy using particle-based epidemic simulation. In Proceedings of the Annual International Conference of the IEEE Engineering in Medicine and Biology Society, Mexico, online, 1–5 November 2021; pp. 1985–1988. [CrossRef]
90. Bayraktar, Y.; Ozyilmaz, A.; Toprak, M.; Isik, E.; Buyukakin, F.; Olgun, M.F. Role of the health system in combating COVID-19: Cross-section analysis and artificial neural network simulation for 124 country cases. *Soc. Work Public Health* **2021**, *36*, 178–193. [CrossRef] [PubMed]

Article

Modeling the Meshing Procedure of the External Gear Fuel Pump Using a CFD Tool

Ihor Romanenko *, Yevhen Martseniuk * and Oleksandr Bilohub

Aircraft Engines Design Department, N. E. Zhukovsky National Aerospace University "Kharkiv Aviation Institute", 61070 Kharkiv, Ukraine; av.belogub@gmail.com
* Correspondence: romanenko.ris@gmail.com (I.R.); y.martseniuk@khai.edu (Y.M.)

Abstract: In modern aircraft engine technology, there is a tendency to replace the mechanical drive of external gear fuel pumps with an electric one. This significantly reduces the integral energy consumption for pumping fuel (kerosene). On the other hand, in order to reduce the dimensions of the structure, it is reasonable to increase the rotation speed of the pumping unit gears. The above considerations make it advisable to study the problems that may arise in the design of pumping units. Analysis of the existing designs of external gear fuel pumps shows that the flow processes in the meshing zone have a significant impact on the pump performance and lifetime. Incorrect truss plate geometry and the compensation system lead to an increase in the velocities when opening and closing the cavity in the meshing zone, which causes intense cavitation. To understand the causes and factors which influence this phenomenon, it is necessary to study the fluid flow behavior in the meshing zone gaps. High-speed cameras are used to experimentally study the flow behavior. However, this approach gives only a qualitative result but does not allow for determining the absolute values of pressure and load in terms of the angle of rotation. Nevertheless, high-speed surveying can be used as a basis for fluid flow model verification. In this paper, the model of the fluid flow in a high-pressure external gear pump was proposed. The verification of the simulation results for HDZ 46 HLP 68 oil operation was carried out according to the results of experimental data visualization. The influence of rotation speed on the position of cavitation zones was revealed and confirmed by operational data. The analysis of the flow process in meshing for kerosene as a working fluid was carried out.

Keywords: gear pump; oil; fuel; fluid dynamics; cavitation; aircraft engine

Citation: Romanenko, I.; Martseniuk, Y.; Bilohub, O. Modeling the Meshing Procedure of the External Gear Fuel Pump Using a CFD Tool. *Computation* **2022**, *10*, 114. https://doi.org/10.3390/computation10070114

Academic Editors: Markus Kraft and Ali Cemal Benim

Received: 13 May 2022
Accepted: 22 June 2022
Published: 6 July 2022

Publisher's Note: MDPI stays neutral with regard to jurisdictional claims in published maps and institutional affiliations.

Copyright: © 2022 by the authors. Licensee MDPI, Basel, Switzerland. This article is an open access article distributed under the terms and conditions of the Creative Commons Attribution (CC BY) license (https://creativecommons.org/licenses/by/4.0/).

1. Introduction

High demand for longevity and lowering costs, in the modern aircraft engine technology, have created new challenges for engineers in the field of engine development; it is necessary to investigate previously studied units under these new working conditions. For fuel systems, these developments led to a move away from the variable-stroke plunger pump, used as the main fuel pump, to the use of an external gear pump with a constant displacement mechanism. The use of a variable-stroke plunger pump allows for the adjustment of the chamber volume and matches the pump performance with the required fluid flow over the entire range of operating modes. The volume of the chamber in the external gear pump is constant and the fluid flow rate depends on rotation speed only. Such pump should be designed for a critical mode. Therefore, it's design becomes significantly oversized for all other operating conditions. Such a pump and engine can be matched over the entire operating range in various ways. The most common method is by using a bypass valve to match the pump capacity to the required engine flow rate. As a result, the integral efficiency of the system is reduced. For more effective matching, it is necessary to remove the dependence of the pump speed on the gas turbine engine (GTE) rotor, which is made possible by using an independent (electric) drive instead of the drive box [1].

In modern aircraft engine technology, there is a tendency to replace the mechanical drive of the external gear fuel pump with an electric one. Transitioning to gearbox-free engines will solve a number of issues associated with the linkage to the rotor speed of the gas turbine engine. This solution also makes it possible to reduce the size of the power plant as well as drag, pipeline numbers, and shaft lines, and to achieve an optimally designed system consistent for the whole operating range. The example of using a gearbox-free turbofan engine is given in the article [2], where the authors showed the layout and advantages of a distributed pump system, confirmed by experimental data.

At the same time, using an electric drive will fix a number of issues associated with the specificity of the journal bearings (Figure 1), used on highly loaded fuel gear pumps. The low viscosity of the working fluid (kerosene), as well as the low peripheral speed of the gear shaft, cause the bearings to operate in a semi-dry friction mode, which significantly accelerates the wear of the bearing surfaces and reduces the life of the pump as a whole. However, applying an electric drive will help to increase peripheral speed up to oil film appearance mode. It will reduce wear on the bearings and consequently increase the life span of the pump.

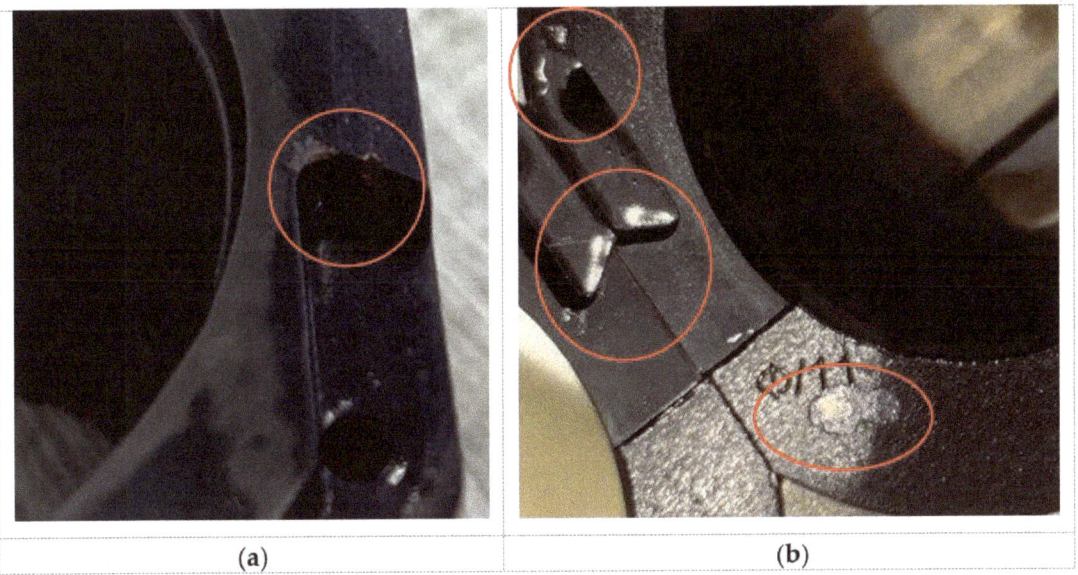

Figure 1. Traces of cavitation wear of the thrust bearing in the meshing area: (**a**) outlet area; (**b**) intel and outlet area.

The rotation speed has both positive and negative effects on the system. With an increase in the rotation speed of the gears, the fluid flow rate, and the fluid displacement rate in the meshing area increase. This will lead to an intensification of the cavitation process and, as a result, to excessive wear of the working chamber. Figure 1 shows the cavitation wear of the thrust bearing surface, which appeared as a result of increasing rotor speed in the external gear fuel pump.

The figure clearly shows traces of wear on the thrust pads in the meshing area. This pump has been designed according to design methodologies for low gear rotation speeds, around 2000–4000 rpm. As a result, its use at higher rotation speeds (around 6000–8000 rpm) led to the occurrence and development of cavitation and damage to the surface of the thrust bearing. These facts became the basis for studying the working process in the meshing zone and assessing the possibility of increasing the gear rotation speed up to 10,000–20,000 rpm and adjusting the design methodology.

The fluid flow behavior in the chambers of an external gear pump can be studied in two ways: by mathematical models simulations and/or by using experimental ways. Modeling the workflow of the gear pump includes determining the operating parameters (velocity and pressure distribution) over the pump chambers. Several approaches exist today [3] for external gear pump modeling but the most commonly used are computational fluid dynamics (CFD) and lumped parameters models. Results of simulation using modeling with lumped parameters for external gear pump have been shown in article [4–9]. This approach uses the division of pump volume for control volumes in which fluid properties are assumed uniform and dependent on time only. Thus, lumped parameter models cannot be used in the current investigation, as they do not take into account fluid flow behavior in the meshing zone in every chamber.

Computer fluid dynamics will provide more detailed modeling with velocity distribution in pump chambers for different gear positions. However, as with a prior model, this approach relies on assumptions depending on the goals of the research. To investigate the circular distribution of pressure and velocity, Szwemin, Campo, and Honga [10–13] used 2D models. This approach clearly showed the distribution of pressure for different positions and rotation velocity but did not sufficiently show the behavior of flow in the inlet and outlet chambers. Yoon [14] used 3D models to study flow behavior. This article shows the pressure distribution in the circumferential direction for high-speed external gear fuel pump, as well as the accuracy of flow rate calculation compared with experimental data. Furthermore, this study has been done for a small-size gear pump, which does not have issues with cavitation in the meshing zone area and is less susceptible to wear and tear. In the article, Corvaglia [14] showed a 3D simulation analysis result for the external gear oil pump. As a result, the distribution of pressure in the circumferential direction for different gears eccentricity was shown in comparison with experimental data. The gas fraction volume and pressure field were obtained for 2000 rpm.

Research using an experimental method is complicated by the peculiarities of the design and working process of gear pumps. Experimentally, it is possible to investigate the pressure distribution around the circumference of the working chamber and the pressure in the cavities of the face-pressure system [15,16], but not in the meshing zone.

Most of the listed researchers have experimentally proved the pressure distribution for the average chamber value, but the current study requires the determination of a more precise behavior in the meshing zone.

Analysis of existing designs shows that the process that occurs in the meshing zone has the greatest impact on the pump operation [17], especially if the design of a closed volume-displacing system is inaccurate. This is caused by the appearance of areas of intense cavitation due to high velocities when closing and opening the closed volume in the meshing zone.

High-speed video (movie) filming to record the nature of the fluid flow can be an informative method to study the behavior of the fluid in the meshing zone. This approach does not allow us to determine absolute values of pressure and forces at a specific moment in time, but it allows us to verify the CFD model of fluid flow for further extrapolation to other sizes and operating conditions. The method and results of such a study were described in [18]. To implement the study, an external gear pump with a transparent housing cover was made, Figure 2.

The results of the experiment are shown as a video of the fluid motion in the gear meshing zone. The video detects the dynamics of the cavitation zone appearance at different operating modes (in terms of speed, inlet pressure, and working fluid temperature). Therefore, in order to verify the calculated CFD model of a high-pressure fuel pump, it was proposed to perform a CFD calculation of the model based on a low-pressure oil pump and to compare the visualization of the experimental and simulation results.

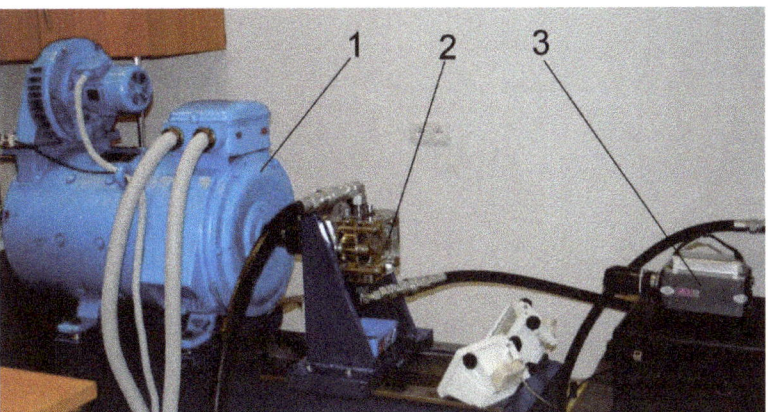

Figure 2. Experimental pump [18].

The purpose of this study is to verify the model of fluid flow in the meshing zone based on the oil pump for further transition to the fuel pump model and analysis of the working process in it. The achievement of this goal is divided into several main stages:
- Preparation of the geometric model and calculation of the geometric parameters of the pump;
- Building a computational grid to perform the calculation of the fluid dynamics in the working cavity;
- Setting up the solver and selecting the turbulence model;
- Analysis of the results and comparison with experimental data;
- Perform the calculation for the fuel external gear pump.

2. Geometric and Calculation Models

Construction of the geometric model is performed according to the parameters of the experimental sample [18]. To solve the problem, it is assumed that the processes occurring in the meshing zone and in the working cavity are uniformly distributed over the height of the gear (boundary effects are not considered). This allowed considering the model in 2-dimensional formulation, simplifying the computational model, and reducing the computational power required. Figure 3 shows the meridional cross-section of the main elements of the pump working cavity, considered in this study.

The model consists of a drive and driven gears, pump inlet (suction), pump outlet (supply), and housing. The construction of a geometric model of a gear pump has a specific trait: the workflow in the pump unit and the model, are affected by the deviation of the geometric dimensions of the pump. To obtain a valid calculation model of the pump, it is necessary to measure the manufactured pump and use the obtained dimensions to construct the actual pump geometry. However, there is no possibility of measuring the experimental pump, so the construction of the geometry of pumping unit elements is made in the middle of the tolerance field of a similar pump size. The relative positioning of the gears and clearances between the housing and the gears are chosen in the middle of the tolerance field (Table 1). The minimum clearance in the meshing is set as the minimum possible, which does not affect the result of the calculation and allows to perform the calculation with the available computing power.

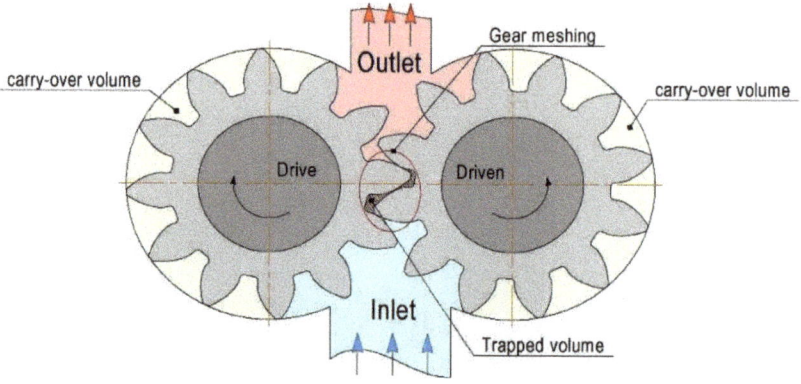

Figure 3. Meridional cross-section of the pump [17].

Table 1. Simulation models' parameters.

Parameters	Units	Value
module (m)	-	5
number of teeth (z)	-	11
housing diameter	mm	69.6
outer gear diameter	mm	69.4
oil density	kg/m^3	880
viscosity	kg/m s	7.5×10^{-3}
inlet chamber pressure	MPa	0.1
outlet chamber pressure	MPa	0.5
gears rotation speed	rpm	1000

Preparing the computational domain is an important step in CFD modeling. In such tasks, the question of the mesh quality, or rather the size of the element is crucial because the calculation is performed in small gaps of the order of 0.1 mm.

For this purpose, the construction of the mesh was implemented considering the specificity of the pump geometry. The mesh in the moving zone was set on the assumption that there should be three elements in the meshing zone, within the gap [19]. Element size outside the meshing zone is increased.

Triangular elements are chosen to describe the computational domain. The size of the elements in the meshing area and on the teeth surface is set to 0.01 mm, in the input and output areas the size of the elements is 1 mm. The results of meshing the computational domain are shown below (Figure 4). Typically, a sensitivity analysis is performed on the parameters of interest but the current study used fluid behavior as the main result. Sensitivity analysis of mesh size cannot be performed because we cannot say which value we expect. That is why the minimal gap value was chosen based on computation capacity.

Figure 4. Meshed pump model.

The Stress Blending (SBE)/Shielded DES turbulence model was selected for the calculation. In article [20], the authors give a detailed description and comparison of the chosen turbulence model with others. They show that this model allows us to describe the work process (workflow) with the optimal ratio of accuracy and computing power consumption. The simulation has been done in a transient mode [21]. Time steps for different liquid types and rotational speeds were around 4×10^{-5} s–4×10^{-7} s. Simulation steps were chosen based on dynamic mesh update and results convergence. Higher values of simulation steps lead to distortion of the shape of the elements and mesh quality which, in a few steps, lead to calculation errors. The max value of y+ on the meshing zone for the oil pump equals 10.

3. Results

The fluid flow analysis divides the meshing process into five characteristic gear positions:

- **Position (a):** start of the drive and driven gear teeth meshing;
- **Position (b):** closed volume appears;
- **Position (c):** compression and reaching a minimum enclosed volume;
- **Position (d):** closed volume increase;
- **Position (e):** closed volume opening.

Figure 5 presents the result for position (a). In this position, the pair of teeth (k_1-k_2) is in the meshing zone and moves toward the suction zone, while the second (m_1-m_2) is approaching the meshing zone.

A gap is formed between teeth k_1 and m_1, and the closing point is the point A_K, which separates the high and low-pressure zones. A meshing teeth pair (m_1 and m_2) are displaced liquid from the interdental chamber. The fluid flow is moved into the high-pressure chamber as an O flow. The pressure in the chamber (k_1-m_1) is higher than the supply pressure. In the chamber (k_2-m_2) the pressure is also increased due to the overflow of liquid from the chamber (k_1-m_1) through the gap J. The pressure increases in the chamber (k_1-m_1) are due to the displacement of the liquid into the high-pressure zone because the hydraulic resistance of the channel between (m_1-m_2) is less than in the gap J.

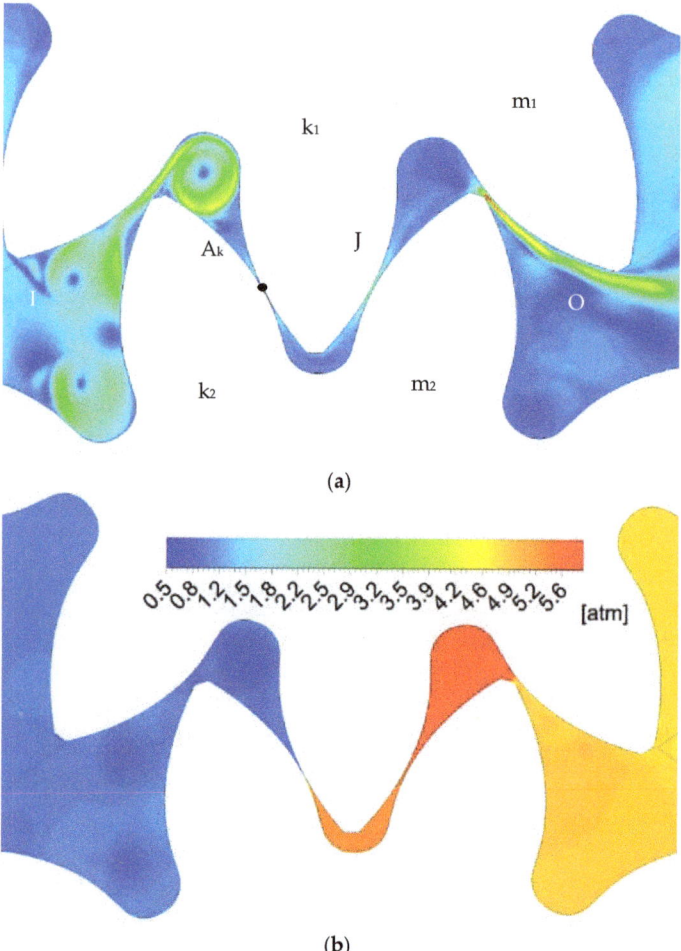

Figure 5. Position a: experimental result [18] picture 3.8; (**a**) velocity field (CFD); (**b**) pressure field (CFD).

Figure 6 presents the result for position (b). In this position, the teeth pair (k_1-k_2) is in the meshing zone and moves into the separation zone.

The second pair of teeth are meshing in A_m point. A closed volume appears and is in a maximum upper position. The A_K point separates the closed volume from the suction chamber and A_m—from the supply zone. The pressure in chambers (k_1-m_1) and (k_2-m_2) is reduced by increasing the volume of the chamber (k_2-m_2) but no negative pressure zone is formed due to the overflow of liquid from the chamber (k_1-m_1) through the gap J. There is a significant pressure drop in the gap J, which is caused by an increase in the fluid flow rate.

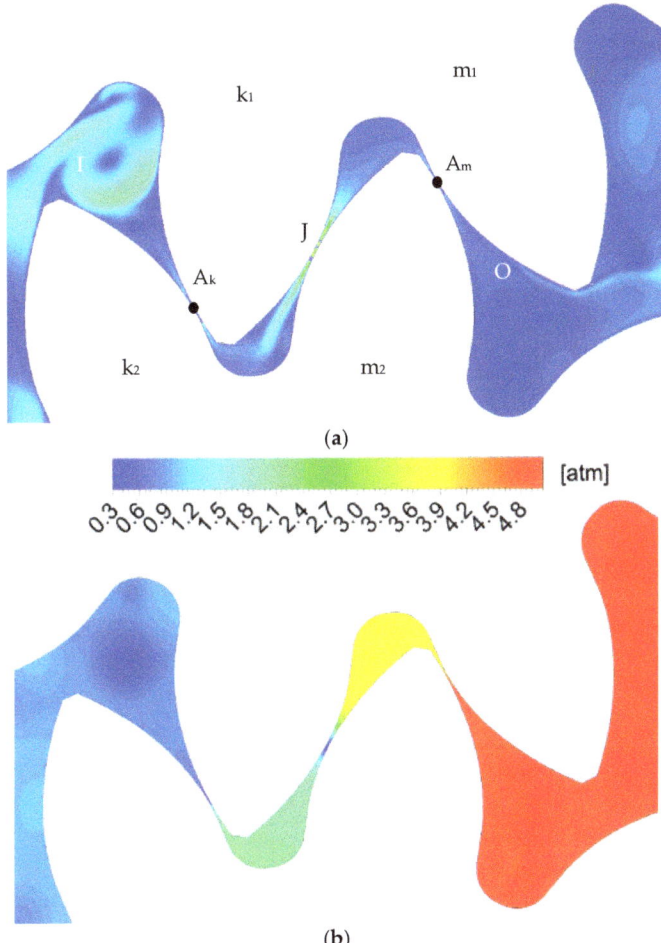

Figure 6. Position b: experimental result [18] picture 3.9; (**a**) velocity field (CFD); (**b**) pressure field (CFD).

Figure 7 presents the result for position (c). In this position, the closed volume is in the center of the meshing zone. The pressure drop between the closed volume and the suction area can lead to fluid over-flow if the thrust bearing gap compensation system is not properly designed. The enclosed volume is separated from the supply zone at the A_m point, and from the suction zone at the A_K point. The fluid flow through gap J from the chamber (k_1-m_1) to (k_2-m_2) increases due to the increase in volume (k_2-m_2) and a significant decrease in pressure inside it. The velocity in the gap J increases even more, which leads to the cavitation zone.

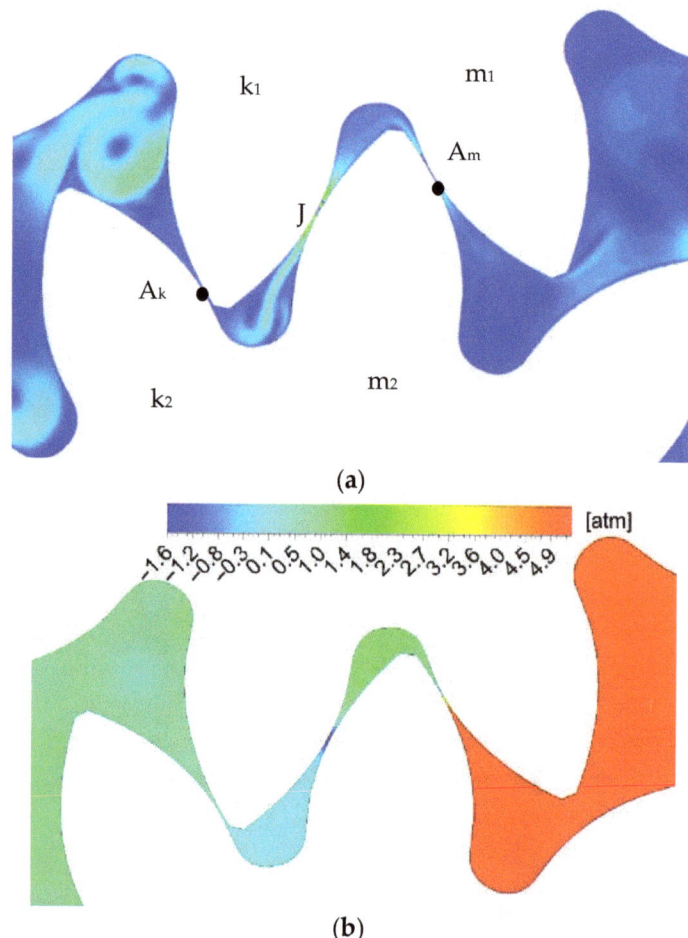

Figure 7. Position c: (**a**) experimental result [18] picture 3.10; (**a**) velocity field (CFD); (**b**) pressure field (CFD).

Figure 8 presents the result for position (d). In this position, a pair of teeth (k_1-k_2) leaves the meshing area, and the second pair is still in the meshing area. The closed volume is in the leftmost position, and its value again takes the maximum value. The pressure in the chamber decreases. The contact point and flow pattern remain the same as in the previous position.

Figure 9 presents the result for position (e). The pair of teeth (k_1-k_2) is out of mesh, and the pair (m_1-m_2) is in contact at the point A_m. Flow I fills the chamber (k_2-m_2) smoothing the pressure to the level of the suction zone. On the other side of the chamber at point A_m there is compression in the supply chamber and the occurrence of fluid flow O. The chamber pressure and fluid behavior for the next pair of teeth are similar to position (a).

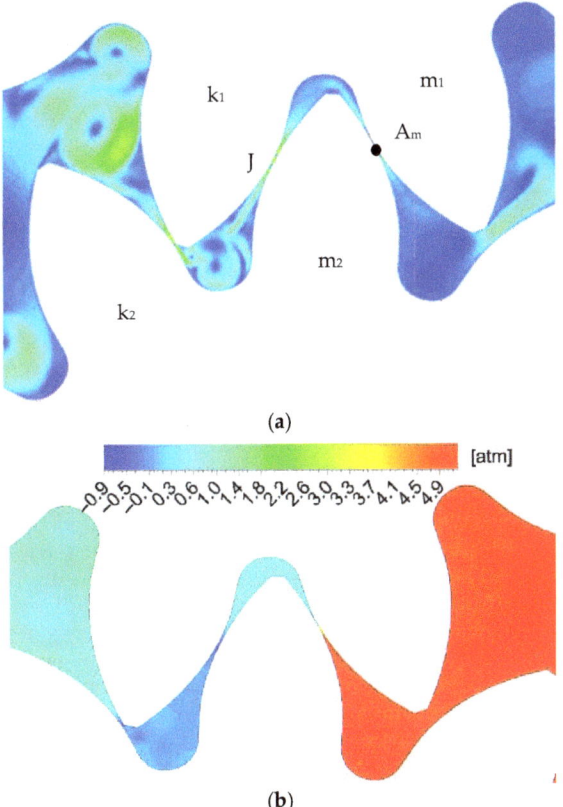

Figure 8. Position d: experimental result [18] picture 3.11; (**a**) velocity field (CFD); (**b**) pressure field (CFD).

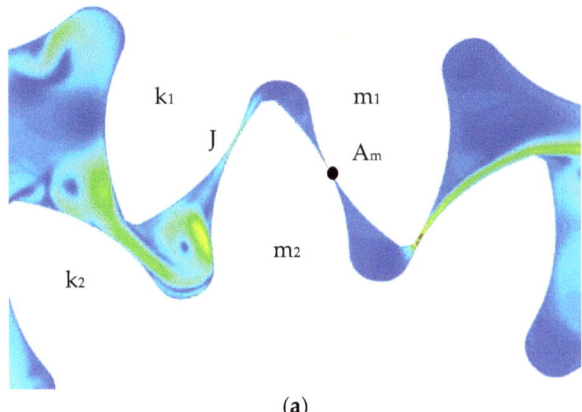

Figure 9. *Cont.*

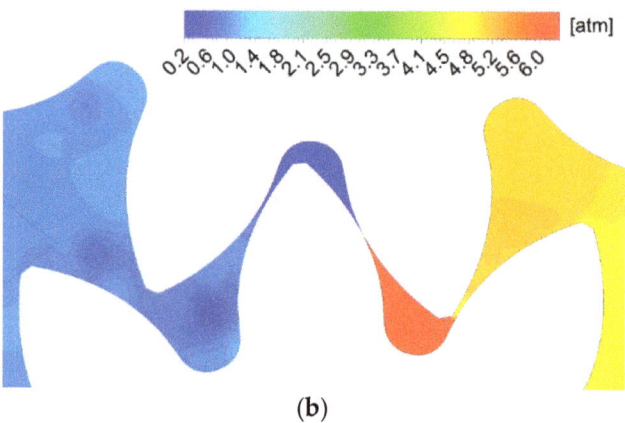

(b)

Figure 9. Position e: experimental result [18] picture 3.12; (**a**) velocity field (CFD); (**b**) pressure field (CFD).

4. Discussion

Experimental data and the numerical experiment showed a similar result and clarified the fluid flow behavior in the meshing zone in more detail. The model can be simplified to two isolated volumes to analyze the effect of the dynamic component of fluid motion. By removing the dynamic component of fluid motion, at the moment of closed volume formation, the pressure in the chamber should correspond to the discharge pressure. This position corresponds to the teeth position (b). Additionally, up to this position, the pressure in the chambers up to the A_K point must correspond to the supply pressure. After the formation of a closed volume, the pressure in the chamber (k_1-m_1) should begin to drop significantly due to an increase in volume, and in (k_2-m_2) to increase due to a decrease in volume. Such fluid motion should continue until the closed volume is opened and connected to the low-pressure zone. Then, the pressure should equalize and reach the level of the suction zone.

Adding the dynamics of fluid motion in the gaps, the fluid flow from the chamber (k_1-m_1) to (k_2-m_2), results in a completely different picture of pressure changes in the meshing area. To analyze pressure changes in the meshing zone, it is necessary to consider chambers (k_1-m_1) and (k_2-m_2) as two sequentially connected closed volumes, connected by a channel corresponding to the gap **J**. Then, after the formation of a closed volume, when the volume of the chamber decreases (k_2-m_2), the pressure will not increase, but due to the flow of liquid through the gap **J** will decrease, increasing the pressure in the chamber (k_1-m_1) up to the position d. This effect changes the position of the low-pressure zone starts and will depend on the gap between the teeth, and therefore on the accuracy of the manufacture of the gear wheel. Comparing the behavior of the fluid in a closed volume with and without the dynamics of fluid motion, one can observe the opposite dynamics of pressure. When the volume in the chamber (k_1-m_1) decreases, the pressure should increase, but since chambers (k_1-m_1) and (k_2-m_2) are connected through gap J, the pressure in (k_1-m_1) decreases because the total volume increases (Figure 10)

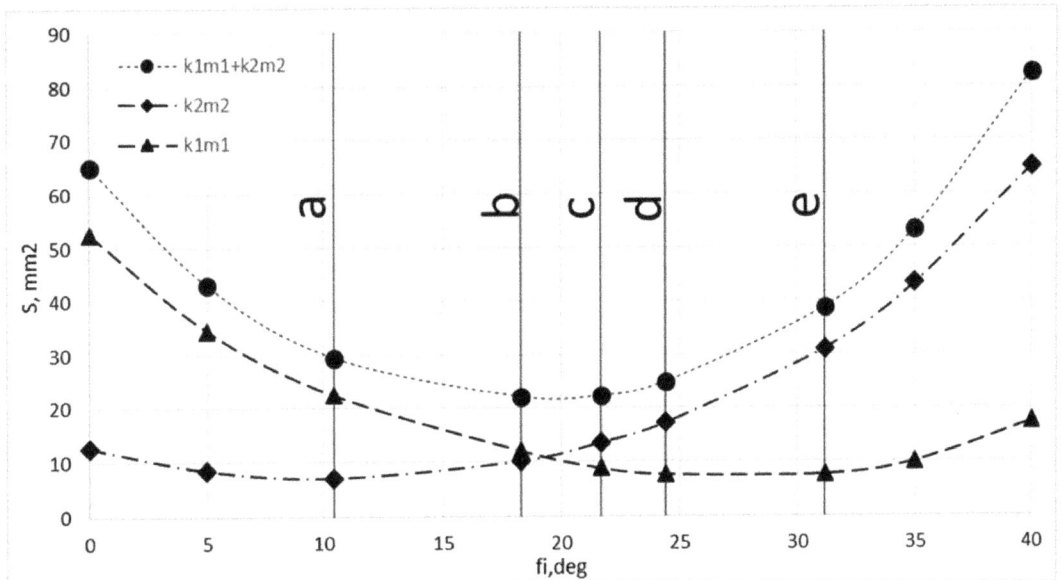

Figure 10. Chamber area dependency in the meshing zone of rotation angle.

This fluid behavior will continue up to position (d) where the chamber (k_2-m_2) connects to the suction zone. In this position, there is also fluid overflow from the chamber (k_1-m_1) to (k_2-m_2) and a slight fluid overflow from the suction area to the chamber (k_2-m_2).

In the model without fluid flow dynamics, in position (e), the pressure in chambers (k_1-m_1) and (k_2-m_2) must correspond to the suction zone pressure, but the pressure in the chamber (k_1-m_1) is lower, which confirms the significant effect of the dynamic effect.

Due to the dynamic effect, the section with the maximum pressure is near position (a) but not in the position with the minimum total volume (Figure 10). This is caused by the fact that as the fluid is displaced from the chamber, the gap between the teeth decreases, but it is still larger than the gap J. To displace the fluid into the high-pressure chamber the fluid to be displaced must have a higher pressure, and as a result, the pressure rises in the chamber until it is sufficient to displace the volume from the chamber along the path of least resistance. The pressure in the chamber increases until the gap is less than J and then the liquid is displaced into the cavity (k_2-m_2).

In order to demonstrate the dependence of pressure maximum and minimum changes on the dynamics of fluid flow, simulations were performed at several gear rotation speeds.

The calculation result (Figure 11) confirmed that increasing the speed will significantly increase the pressure in the zone with the maximum pressure, due to the effects described above, and in the zone with the minimum pressure will significantly reduce it, while increasing the cavitation zone.

According to the goals set in this study, the simulation of fluid flow in the external gear pump operating on kerosene was performed. Table 2 presents the geometric and operating parameters for this study.

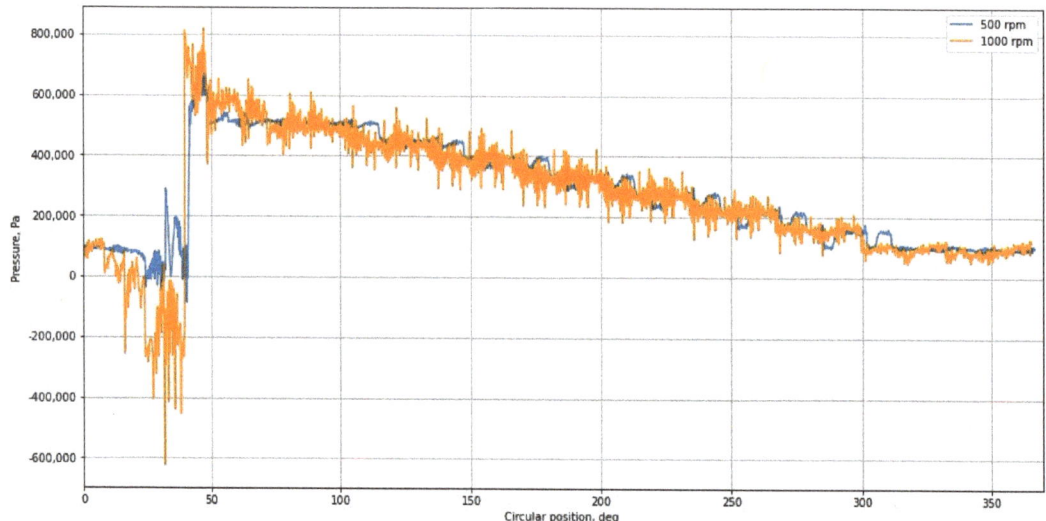

Figure 11. Pressure changes in circumferential direction for 500 rpm and 1000 rpm.

Table 2. Simulation model parameters for the fuel system.

Parameters	Units	Value
module (m)	-	5
number of teeth (z)	-	11
housing diameter	mm	61.0
outer gear diameter	mm	60.84
fuel density	kg/m^3	780
viscosity	kg/m s	2.4×10^{-3}
inlet chamber pressure	MPa	0.5
outlet chamber pressure	MPa	10
gears rotation speed	rpm	8000

For the simulation of the fuel pump, the same rules and assumptions were established as for the oil one. Pressure and velocity fields for all previously introduced gear positions (a–e) are shown below in Figures 12 and 13.

Unlike oil, the results of fuel calculation show a significant influence of fluid flow dynamics on the pressure field in the entire working cavity. The sections in the meshing zone show a much greater pressure drop. This corresponds to the resulting much higher velocity that is the result of two factors: a significant increase in rotation speeds up to 8000 rpm (in eight times) and a decrease in fluid viscosity by three times. Compared to the calculation of the oil pump in position (a), the vertexing of the flow is larger. The occurrence of low-pressure zones passes from position (c) to position (b). In positions (c) and (e), the negative effect of the dynamic component of fluid movement is much stronger than in the oil pump: the volume of the area with increased pressure and the area with decreased pressure increases significantly. This fluid behavior will lead to unpredictable behavior of the end gap compensation system and cavitation wear in the suction area.

The results of the fuel pump simulation were compared with the traces of cavitation wear on the end bearings of the real pump (Figure 14). Wear traces are located in areas with minimum pressure in the suction zone, as well as in areas of vortex formation. This fact also confirms the validity of the model used.

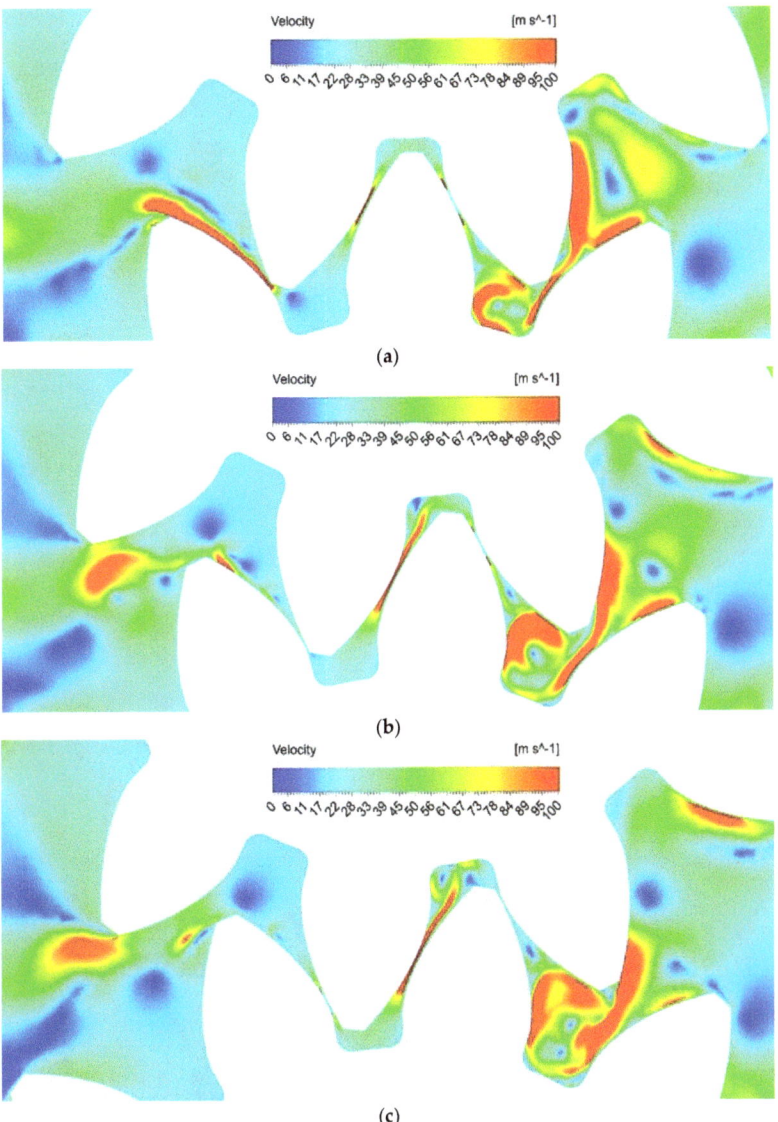

Figure 12. Cont.

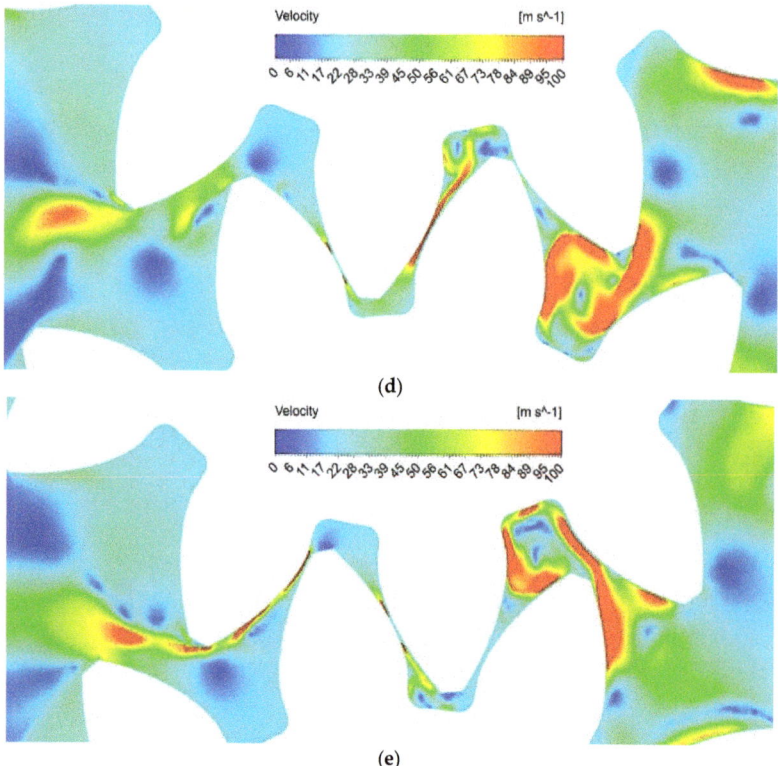

Figure 12. Velocity distribution for fuel pump ((**a–e**) in accordance with the position in Figure 10) a = 0.0002 s, b = 0.00038 s, c = 0.00045 s, d = 0.0005 s, e = 0.00065 s.

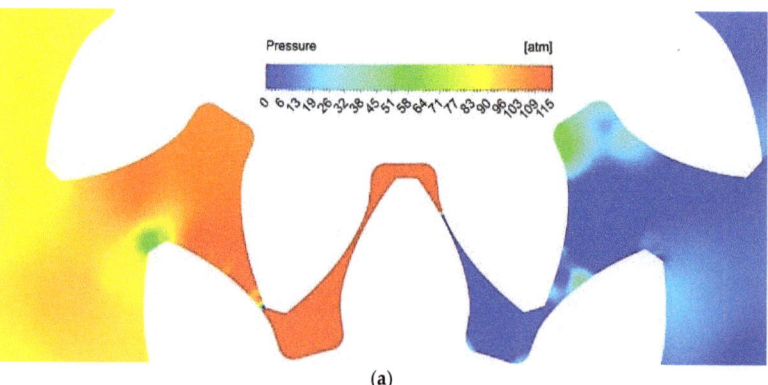

Figure 13. *Cont.*

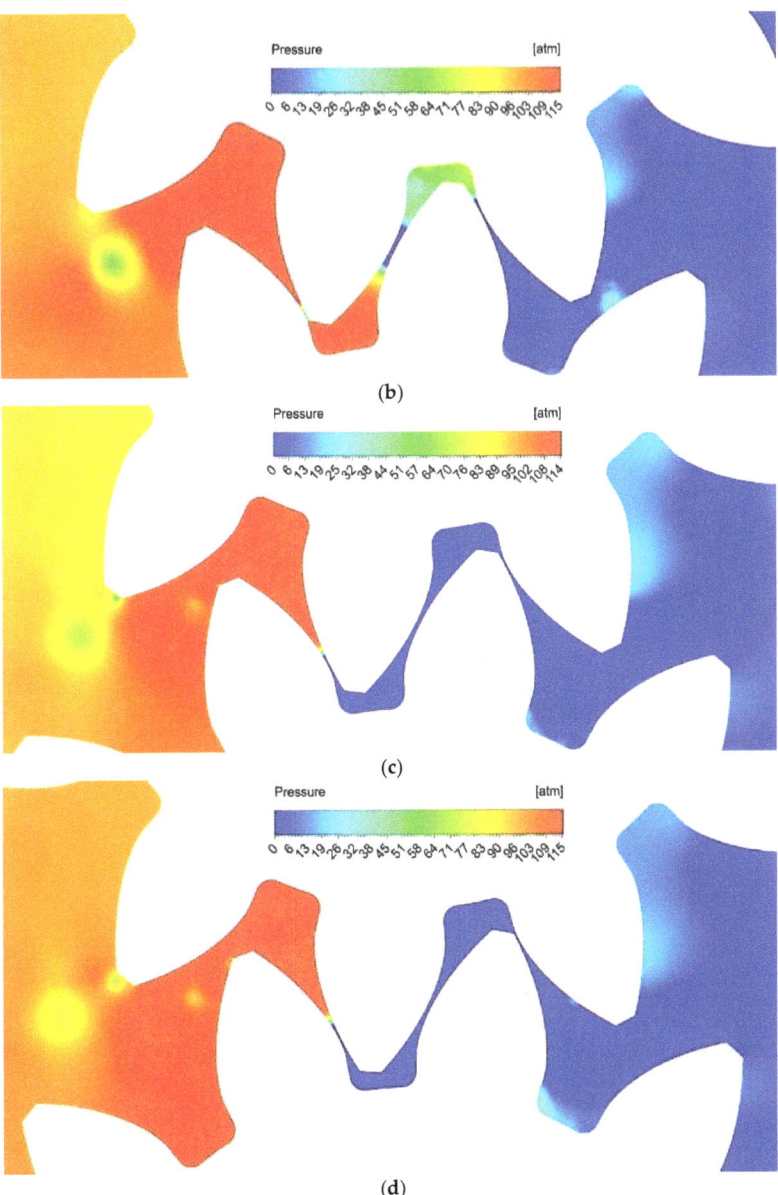

Figure 13. *Cont.*

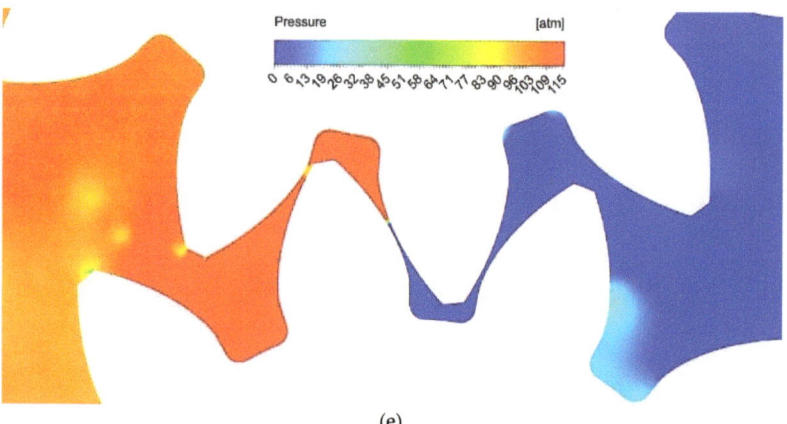

(e)

Figure 13. Pressure distribution for a fuel pump ((**a**–**e**) in accordance with the position in Figure 10) a = 0.0002 s, b = 0.00038 s, c = 0.00045 s, d = 0.0005 s, e = 0.00065 s.

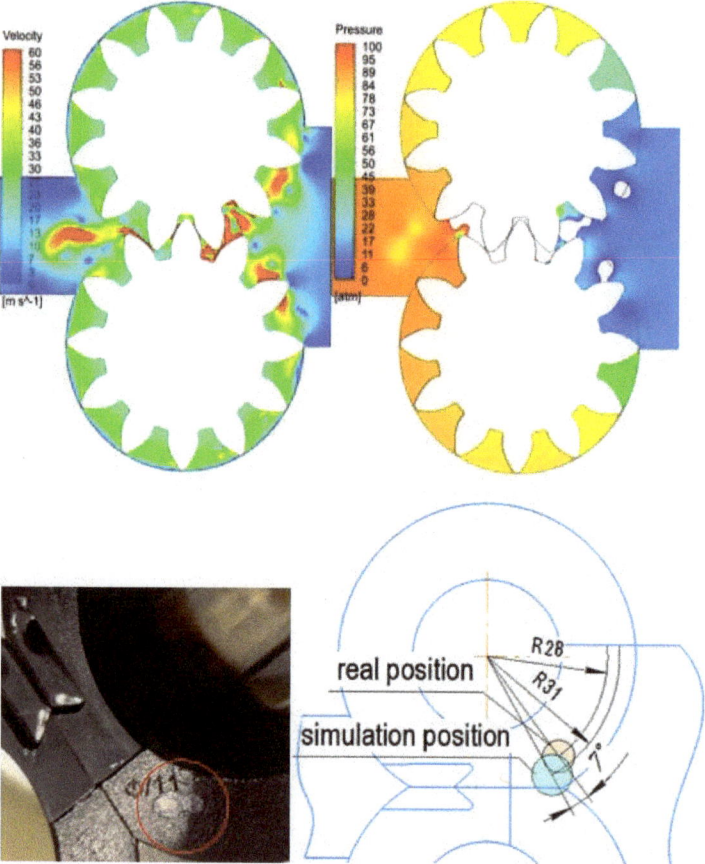

Figure 14. Comparison of the damage to the thrust bearing and simulation results.

5. Conclusions

1. Based on visualization analysis, the characteristic positions of the gears were determined in which there is a change in the nature of the flow of fluid due to changes in the geometry and volume of chambers. The appearance of the vortex between k_1 and m_2 begins when the closed volume (c) is closed and intensifies significantly after the opening of the closed cavity. The vortex attenuates in the suction chamber area after the breakaway from the tooth, closer to the end of the suction zone (Figure 14).
2. Comparison of the frame-by-frame high-speed shooting of the oil pump operation and numerical simulation showed that all characteristic positions of the gear are similar in the nature of the flow and formation of vortices.
3. The positions of the gears at which the pressure reaches the maximum (a) and minimum (c) values have been determined.
4. The influence of fluid flow velocity on global maxima and minima of pressure in the working cavity is revealed. When rotation velocity is increased in the suction zone chambers, the pressure decreases, in the supply zone chambers the pressure increases.
5. Comparison of damages of thrust bearings with the characteristic formation of vortices and zones of low pressure is performed. The cavitation damage is located at a radius of 28 mm, and according to the simulation results, the discharge zone is located at a radius of 31 mm. The displacement in a circumferential direction between the real design and the simulation results is 7 degrees.

The conducted study allowed us to clarify the behavior of the fluid in the meshing zone, to determine the influencing factors on the gap end compensation system to further improve the design and service life of the rocker unit and transition to higher rotational speeds.

Author Contributions: Conceptualization, O.B. and I.R.; Formal analysis, Y.M.; Investigation, I.R.; Methodology, I.R. and O.B.; Project administration, O.B.; Resources, I.R.; Software, I.R.; Supervision, O.B.; Validation, I.R.; Visualization, I.R.; Writing—original draft, I.R.; Writing—review & editing, Y.M. and O.B. All authors have read and agreed to the published version of the manuscript.

Funding: This research received no external funding.

Institutional Review Board Statement: Not applicable.

Informed Consent Statement: Not applicable.

Data Availability Statement: Not applicable.

Conflicts of Interest: The authors declare no conflict of interest.

References

1. Tom, L.; Khowja, M.; Vakil, G.; Gerada, C. Commercial Aircraft Electrification—Current State and Future Scope. *Energies* **2021**, *14*, 8381. [CrossRef]
2. Popov, V.; Yepifanov, S.; Kononykhyn, Y.; Tsaglov, A. Architecture of Distributed Control System for Gearbox-Free More Electric Turbofan Engine. *Aerospace* **2021**, *8*, 316. [CrossRef]
3. Bilohub, O.; Romanenko, I.; Hudoshnyk, O.; Trystan, S. The performance capabilities of the journal bearings as a supports of the fuel gear pump of the gas turbine engine. *Aerosp. Tech. Technol.* **2021**, *2*, 52–58. [CrossRef]
4. Rundo, M. Models for flow rate simulation in gear pumps: A review. *Energies* **2017**, *10*, 1261. [CrossRef]
5. Shah, Y.G.; Vacca, A.; Dabiri, S. Air Release and Cavitation Modeling with a Lumped Parameter Approach Based on the Rayleigh-Plesset Equation: The Case of an External Gear. *Energies* **2018**, *11*, 3472. [CrossRef]
6. Battarra, M.; Mucchi, E. On the assessment of lumped parameter models for gear pump performance prediction. *Simul. Model. Pract. Theory* **2020**, *99*, 102008. [CrossRef]
7. Marinaro, G.; Frosina, E.; Senatore, A. A Numerical Analysis of an Innovative Flow Ripple Reduction Method for External Gear Pumps. *Energies* **2021**, *14*, 471. [CrossRef]
8. Torrent, M.; Gamez-Montero, P.J.; Codina, E. Model of the Floating Bearing Bushing Movement in an External Gear Pump and the Relation to Its Parameterization. *Energies* **2021**, *14*, 8553. [CrossRef]
9. Chen, L.; Zhao, Y.; Zhou, F.; Zhao, J.; Tian, X. Modeling and Simulation of Gear Pumps based on Modelica/MWorks. In Proceedings of the 8th Modelica Conference, Dresden, Germany, 20–22 March 2011.

10. Szwemin, P.; Fiebig, W. The Influence of Radial and Axial Gaps on Volumetric Efficiency of External Gear Pumps. *Energies* **2021**, *14*, 4468. [CrossRef]
11. del Campo, D.; Castilla, R.; Raush, G.; Gamez-Montero, P.J.; Codina, E. Numerical Analysis of External Gear Pumps Including Cavitation. *J. Fluids Eng.* **2012**, *134*, 081105. [CrossRef]
12. Hong, W.; Zhang, K.; Yao, Y. Numerical simulation of the influence of gear rotation speed on the gear pump flow field. *Appl. Mech. Mater.* **2015**, *741*, 232–236. [CrossRef]
13. Houzeaux, G.; Codina, R. A finite element method for the solution of rotary pumps. *Comput. Fluids* **2007**, *36*, 667–679. [CrossRef]
14. Yoon, Y.; Park, B.H.; Shim, J.; Han, Y.O.; Hong, B.J.; Yun, S.H. Numerical simulation of three-dimensional external gear pump using immersed solid method. *Appl. Therm. Eng.* **2017**, *118*, 539–550. [CrossRef]
15. Corvaglia, A.; Rundo, M.; Casoli, P.; Lettini, A. Evaluation of Tooth Space Pressure and Incomplete Filling in External Gear Pumps by Means of Three-Dimensional CFD Simulations. *Energies* **2021**, *14*, 342. [CrossRef]
16. Zhoua, J.; Vaccab, A.; Casoli, P. A novel approach for predicting the operation external gear pumps under cavitation condition. *Simul. Model. Pract. Theory* **2014**, *45*, 35–49. [CrossRef]
17. Romanenko, I.; Bilohub, O. Fuel gear pump of gas turbine engine operating processes simulation issues analysis. *Aero-Space Tech. Technol.* **2020**, *2*, 24–30. [CrossRef]
18. Kostiuk, D.V. Increasing Efficiency of the Gear Pump by Reducing the Intensity of Cavitation Phenomena in Its Working Cavities. Ph.D. Dissertation, National Technical University of Ukraine "Igor Sikorsky Kyiv Polytechnic Institute", Kyiv, Ukraine, 17 May 2017.
19. Castilla, R.; Gamez-Montero, P.J.; Erturk, N.; Vernet, A.; Coussirat, M.; Codina, E. Numerical simulation of turbulent flow in the suction chamber of a gearpump using deforming mesh and mesh replacement. *Int. J. Mech. Sci.* **2010**, *52*, 1334–1342. [CrossRef]
20. Menter, F.; Hüppe, A.; Matyushenko, A.; Kolmogorov, D. An Overview of Hybrid RANS–LES Models Developed for Industrial CFD. *Appl. Sci.* **2021**, *11*, 2459. [CrossRef]
21. Fletcher, C.A.J. *Fletcher Computational Techniques for Fluid Dynamics*; Springer: Sydney, Australia, 1991; ISBN 3-540-54304-X.

Article

Crack Identification in Cantilever Beam under Moving Load Using Change in Curvature Shapes

Nutthapong Kunla [1], Thira Jearsiripongkul [1,*], Suraparb Keawsawasvong [2] and Chanachai Thongchom [2]

1. Department of Mechanical Engineering, Faculty of Engineering, Thammasat School of Engineering, Thammasat University, Pathum Thani 12120, Thailand; k.nut610@gmail.com
2. Department of Civil Engineering, Faculty of Engineering, Thammasat School of Engineering, Thammasat University, Pathum Thani 12120, Thailand; ksurapar@engr.tu.ac.th (S.K.); tchanach@engr.tu.ac.th (C.T.)
* Correspondence: jthira@engr.tu.ac.th

Abstract: Cracks in structural components may ultimately lead to failure of the structure if not identified sufficiently early. This paper presents a crack-identification method based on time-domain. Captured time-domain data are processed into central difference approximation of displacement of each node (point) in the structure. Abnormally high central difference approximation of displacement of a node relative to those of its neighborhood points indicates a crack at that point. A suite of simulation experiments and numerical calculations was conducted to find out whether the proposed identification method could accurately identify the location of a crack in a cantilever beam under moving load compared to the location found by an exact solution method, and the outcomes indicated that it was as able as the analytical method. The proposed method is an FEA analysis, an approach familiar to virtually every engineer. Therefore, the relative amount of time and effort spent on developing the proposed method for a specific application is much less than those spent on developing an analytical method. The saved time and effort should enable more engineering personnel to perform routine checks on structural elements of their interest more simply and frequently.

Keywords: crack identification; curvature shape; central difference approximation; cantilever beam; moving load; forced vibrational response

1. Introduction

Methods for identifying damage in structures, such as aircrafts, bridges, ships, and buildings is an important area of research and development that has a great potential for cost-saving and safety improvement to those structures [1–3]. The presence of cracks in a structure brings about local variations in the stiffness of the structure. The extent of such variations mainly depends on the depth and location of the cracks, which affects the dynamical behavior of the whole cracked structure. Vibration-based damage-detection methods based on change in physical properties of a structure have attracted the attention of many developers over the past few decades. Originally, Dimarogonas developed a theory of vibration of cracked shafts and wrote it up in a textbook [4]. Nevertheless, the crack-identification method based on this theory could indicate whether there was a crack or not but could not specify the crack location. Later, a new technique was developed to achieve both, based on a new parameter of change in displacement mode shape of a beam called "Curvature Mode Shape". Pandey et al. [5] was the first to develop a version of it. Their version successfully identified typical crack locations. The study by Pandey et al. [5] was adapted by Sahin and Shenoi [6] to obtain a damage-detection algorithm based on the combined method of global (changes in natural frequencies) and local (curvature mode shape) analysis. A gapped smoothing damage-detection method was also carried out by Ratcliffe and Bagaria [7] for evaluating a damage index of composite beams. A modal curvature-based method was also employed by Hamey et al. [8] to determine cracks in carbon/epoxy composite beams. In addition, Qiao et al. [9] demonstrated the use of dynamic-based damage detection techniques to identify cracks in composite laminated plate. However, those previous studies were not able to identify a crack at some node—a node is a point at which no motion occurs—exists in a beam. Later, to counter the issue of crack identification at the node existing on the beam, two development teams, Chandrashekhar et al. [10] and Frans et al. [11], each proposed a similar method but based on additional pre-selected vibrational modes of a beam. The method was demonstrated to be effective as intended. However, it was not practical for on-site analysis of various structures because not all factories were equipped with sophisticated tools to measure those extra vibrational parameters.

This paper attempted to develop a curvature mode shape method that did not rely on difficult-to-obtain vibrational parameters. The developed method would also be based on moving load because we learned that moving load or moving mass on a bridge produced larger deflections and higher stresses compared to the case when an equivalent load was applied statically, and hence the deflections would be easier to measure precisely [12–16]. This idea of easier detection by using a variety of moving loads in the analysis was inspired by the results of a study by Chouiyakha et al. [17] as well as the results of a study by Roveri and Carcaterra [18] that demonstrated that their proposed methods based on moving load were truly able to identify the locations of the cracks precisely. The approaches based on moving load for damage detection in structures were widely used by several researchers in the past [19–22]. In contrast, our proposed method would be based on vibrational data in time domain instead of vibrational data in frequency domain (used in most of the studies mentioned above) to counter a common problem for many engineers that is responsible for crack identification of their facilities: complex and time-consuming tasks encoding unnecessarily complex mathematical calculation steps into a functioning and precise crack-identification app for their intended structure. The complexity of the mathematical calculation steps for finding a solution based on frequency domain data was higher than that based on time domain, as stated by Asnaashari et al. [23].

In addition, the developed method from the present study should work right out of the box—engineer users would not have to derive any exact analytical expression for the dynamics of the structural components under investigation to use the method; only readily available on-site data would be needed for the method to make accurate predictions.

The rest of the paper is organized as follows. Section 2 describes our methodology of crack-identification scheme. Section 3 reports numerical results and discusses them. Section 4 concludes the paper. Finally, Section 5 suggests future work and limitations of the proposed method.

2. Methods

2.1. A Previous Crack Identification Method

When a crack occurs in the structure of a beam, some structural properties will change. A crack in a beam can be identified by the curvature mode shape of the beam, as presented by Pandey et al. [2]. The authors now introduce a verified relationship between the flexural stiffness and amplitude of curvature of natural modes of vibration at a point in or on a common beam, which could be used to identify a crack effectively.

The curvature of a beam is related to the flexural stiffness of the beam. The curvature at a point in or on the beam is given by Equation (1),

$$\kappa = \frac{M}{EI} \tag{1}$$

where κ is the curvature at that point; M is the bending moment at the section; E is the modulus of elasticity; and I is the second moment of the cross-sectional area. If a crack occurs in a structure, it reduces the EI of the structure at the cracked point, which makes the magnitude of the curvature higher at that point. The change in the curvature is local in nature, and hence it can be used to locate a crack in a small, suspected region.

Pandey et al. [5] used the displacement (or magnitude) of mode shape to calculate the curvature mode shape by central difference approximation as in Equation (2) below,

$$y'' = (y_{i+1} - y_i + y_{i-1})/h^2 \qquad (2)$$

where y'' = the central difference approximation at node or point i; h = the length between node i and $i + 1$; y = the magnitude (or displacement in the original paper) of mode shape at node i; i = node number, $i = 1, 2 \ldots m$; and m = Total number of nodes.

A crack at node i is identified by the absolute difference in the central difference approximation at that node.

$$\Delta y''_i = |y''_{c_i} - y''_{uc_i}| \qquad (3)$$

$\Delta y''_i$ = the absolute difference in the central difference approximation at node i; y''_{c_i} = the central difference approximation of the cracked beam at node i; y''_{uc_i} = the central difference approximation of an uncracked beam at node i.

2.2. Proposed Identification Method

2.2.1. Main Modules in the Proposed Method

The proposed crack-identification method consists of three main modules: dynamic deflection module, central difference approximation calculation module, and crack-identification module. The first module determines the average dynamic deflection (vertical displacement) of every point (node) on the beam. The second module takes this set of deflection values as input and processes them into a set of central difference approximation values of every point on the beam. Finally, this set of central difference approximation values is inputted into the crack-identification module and processed, the output of which will be a graph of magnitude of central difference approximation at every point (node) of the beam. A sharp peak at any point in this graph indicates a crack at that point. All three modules were manually MATLAB-coded by the author. The flow diagram of the operations of these main modules is illustrated in Figure 1 below.

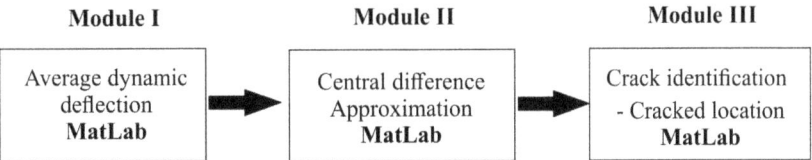

Figure 1. Flow diagram of the operations.

2.2.2. Rationale behind the Proposed Method

The rationale behind the proposed method is that time and effort is hugely saved in composing simulation experiments in an all-familiar FEA framework than composing a physical experiment or manually coding analytical methods to solve a particular problem.

Early on in this research project, the authors attempted to apply the magnitude of mode shape method [4] to identify a cracked beam by themselves, but we encountered two unexpected issues that might also deter other engineers from effectively applying the method to their research work.

Firstly, the frequencies of several natural vibrational modes of an investigated beam had to be obtained from a costly and time-consuming measurement setup using sophisticated and specialized instruments.

Secondly, the operational steps of the curvature mode shape crack-identification method consist of determining the vertical displacement of points (nodes) in and on the beam and using those displacement values to construct curvature mode shapes, then identifying the crack location from the relationship between the flexural stiffness of points in and on the beam and the constructed curvature mode shapes. These operational steps were lengthy and time-consuming, both in program coding effort and computational time. Since the only necessary data point for identifying a crack is the flexural stiffness at that point, and since that point can be determined directly by a basic beam displacement measurement setup without any need for theoretical or experimental determination of natural vibrational modes, it will be simpler, easier and more efficient for engineers to use the proposed method.

In the operations of the proposed method, the authors observe the local peak of the absolute difference in the central difference approximation of the cracked beam and the uncracked beam at a point in or on the beam, obtained from the vertical displacement value of the beam at that point (the vertical displacement was a result of a load moving along the beam). This local peak is then the indicator of a crack. To summarize, a crack is observed as a positive peak at a point on or in a beam of the graph of vertical central difference approximation from vibration due to moving load versus position coordinates.

A moving load or mass induces vibration of the structural element along which it travels. It produces a larger deflection and a higher stress compared than an equivalent static load. The deflection is a function of both time and speed of the moving load.

Getting back to the practical disadvantage of the mode shape method, it is not easy to obtain an accurate magnitude of central difference approximation through differentiation of mode shape. Moreover, differentiation may further amplify measurement error. In contrast, since vertical displacement at a point in and on a beam is easy to obtain by a basic displacement sensor, the central difference approximation in or on every point of a beam can be readily determined by using central difference approximation. When central difference approximation values of displacement of all points are plotted on a graph, it will show a crack at a point (node) where a peak is located.

2.2.3. Detailed Operations in Each Main Module

As shown in Sub Section 2.2.1, the proposed system has three main modules. The operational procedure of each module is described in this subsection.

The first module is the dynamic deflection module, consisting of two steps. In the first step, from a specified velocity of the moving load parameter, external force acting on each specified node —1, 2, 3, ..., m—as shown in Figure 2, this module determines the time step and the total number of time steps for average displacement calculation.

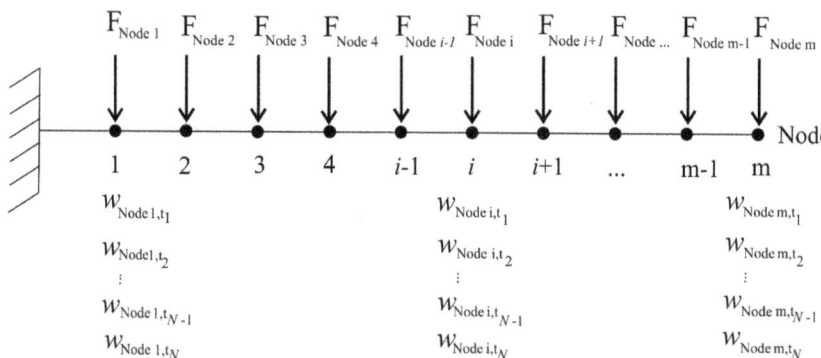

Figure 2. The magnitude of displacement and external force acting on each specified node.

The time of contact T on a beam depends on the speed of the moving load on the beam contact surface. One-time step ΔT is calculated as below,

$$\Delta T = \frac{T}{p \cdot n} \quad (4)$$

where p is a specified number of intervals, and for each interval, n is the number of sub-intervals. the total number of time steps, N, for a specified speed of moving load is evaluated as below,

$$N = p \cdot n. \tag{5}$$

In the second step, the first module calculates the average displacement values of every node. Then, the displacement values at every point for each time step are summed, and the average displacement values at every point for the whole time T is calculated as a sum below,

$$w_i = \frac{1}{N}\sum_{j=1}^{N} y_{i,j} \tag{6}$$

where w_i = the average displacement value at point i for the whole time T; $y_{i,j}$ = the displacement value at point i at time step j. i = Node label, 1, 2 ... , m; j = Number of time steps along which the load moves from one fixed end to the other open-end of the beam, 1, 2 ... , N.

The second module calculates the magnitude of central difference approximation of the average displacement value output obtained from the first module. This magnitude of central difference approximation of average displacement under moving load is calculated with central difference approximation as follows,

$$w_i'' = \left| \frac{1}{N}\sum_{j=1}^{N} y_{i+1,j} - 2\frac{1}{N}\sum_{j=1}^{N} y_{i,j} + \frac{1}{N}\sum_{j=1}^{N} y_{i-1,j} \right| / h^2 \tag{7}$$

where w'' = the central difference approximation of average displacement of each node; $y_{i,j}$ = Displacement at node i at time step j; h = Length between two nodes.

This central difference approximation procedure was coded in MATLAB.

The third module calculates the absolute difference in the magnitude of central difference approximation of displacement between each node, which are parallel to one another on the uncracked beam and cracked beam (Section 2.1). It was also coded in MATLAB.

2.3. Finite Element Analysis of Beam Vibration under Moving Load

FEA analysis was run to evaluate the efficiency of the proposed method, as described in Section 2.2. Displacement results obtained from the FEA approach could be substituted into Equation (7) to obtain a graph of central difference approximation of displacement versus position, in which the location of an abnormal peak would be identified as the crack location. In this subsection, the background of this kind of finite element analysis is described.

Finite Element Analysis (FEA) is a simulation of physical phenomenon with a numerical technique called Finite Element Method. In our proposed method, vibration in or on every point (node) of the beam under moving load needs to be simulated with this numerical technique in combination with the parameters of the static beam. Therefore, the simulation needs to include a moving load to induce vibration.

The governing equation and boundary conditions of a beam vibrating under the influence of a moving load are Equations (8)–(12) below,

$$EI\frac{\partial^4 y(x,t)}{\partial x^4} + \rho A \frac{\partial^2 y(x,t)}{\partial t^2} = F\delta(x - vt) \tag{8}$$

where E is the young's modulus of elasticity of the beam; I is the second moment of inertia of the beam cross section; ρ is the material density of the beam; A is the area of cross section of the beam; x is the point of interest on the beam; v is the speed of load moving along the beam; t is time from the start. A simple beam model labelled with these variables is illustrated in Figure 3 below.

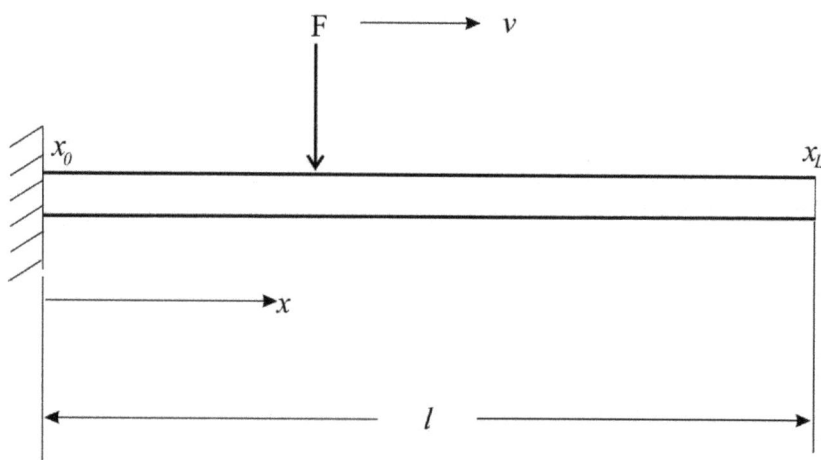

Figure 3. Cantilever beam subjected to a load.

One end of the beam is fixed to a location ($x = 0$), while the other end ($x = l$) can move freely when a force is acting on it, hence

$$y(0,t) = 0 \tag{9}$$

$$\frac{\partial y(0,t)}{\partial x} = 0 \tag{10}$$

$$\frac{\partial^2 y(l,t)}{\partial x^2} = 0 \tag{11}$$

$$\frac{\partial^3 y(l,t)}{\partial x^3} = 0 \tag{12}$$

In an FEA simulation, those equations above are represented by the matrix equation below, Equation (13), reported by Lee [24],

$$[M]\{\ddot{D}\} + [K]\{D\} = \{F\} \tag{13}$$

where $[M]$ is the n × n mass matrix of the discrete system; $[K]$ Is the n × n stiffness matrix of the discrete system; $\{D\}$ Is the n × 1 vector of displacement magnitude of every node in the discrete system; $\{F\}$ Is the n × 1 vector of external forces on every node in the discrete system.

These matrix equations are selectable in the transient structure analysis module of a popular FEA software named ANSYS, and are familiar to every engineer.

The FE settings for the dimensions and material properties of the beam for every simulation run are tabulated in Table 1 below, following exactly those values for the beam in an analytical work of Lin and Chang [25]. The input of the beam was a 3D model of the beam, created by SolidWork software. The dimensions of the beam were already specified in the creation of the 3D model, but the material property settings were specified in the Ansys FEA software [26]. The outputs were values of vertical displacement of every node in a Microsoft Excel file.

Table 1. FE Settings for Dimensions and material properties of the beam.

Material Properties	Value	Unit
Dimension	580 × 20 × 20	mm
Density	7800	kg/m^3
Young's Modulus	206	GPa
Poisson Ratio	0.3	

In detail, a number of FE simulation runs were carried out on a 3D model of a cantilever beam hosting a transverse notch (representing a physical crack). A three-dimensional geometrical model of solid cantilever beam, constructed with SolidWorks software and saved as an *IGES file*, was imported into the ANSYS software.

2.4. Evaluation of the Finite Analysis Model and Transient Simulation

To ensure the suitability of the FE model and the transient simulation of the cracked cantilever beam, we verified them against the results of a study by Lin and Chang [25] of a cantilever beam with the same dimensions and mechanical properties. The verification process against the analytical solution of their investigated beam was done in two approaches: (i) verifying by using natural transverse frequency with Fourier analysis and (ii) verifying by using forced deflection responses at the free end of the cracked cantilever beam under moving load.

The first approach used a model of cracked cantilever beam with a modeled crack at $x_1/l = 0.3$ and a notch depth of 30% of the beam thickness. An impulse load was applied on the free end. The magnitude of this load was 100 N applied for 0.001 s on one node at the free end of the cantilever, as shown in Figure 4. A transient analysis of the beam was conducted in ANSYS to simulate the free vibration response of the beam due to the impulse load. The length of the beam, L, was 580 mm. A Newmark's integration scheme was employed as the solver, and the time increment was kept fixed at 0.000025 s. After the impulse load was applied, the free vibration of the beam was calculated up to 0.1 s. The first three natural frequencies of the beam, calculated by applying fast Fourier transform of acceleration at the mid-span of the beam, were then compared with the analytical values obtained from the closed form solution reported in [25]. As listed in Table 2, the FE results were less than 3% different from the analytical results, indicating that the FE model and the transient analysis were reasonably accurate in simulating the dynamic behavior of the beam.

Table 2. Comparison of natural frequencies calculated from FE model and analytical solution.

Mode	Natural Frequency (Hz)		
	Analytical [25]	FE Model	Percentage Error
1	30.88	30	2.85
2	195.60	200	2.24
3	540.48	540	0.09

The second approach compared the forced deflection responses at the free end of the cracked cantilever beam under moving load (from the proposed method) and those from the analytical method. The speed of the moving load was 0.6 time of the critical speed (V_{crit}). This speed was 0.6 time critical speed V_{crit}, where V_{crit} was defined as $V_{crit} = 1.8751/L(\sqrt{EI/\rho A})$ by Lin and Chang [25]. A speed higher than this critical speed would cause insignificant deflection, while a speed lower than this speed will make the deflection more clearly observable. The forced displacement responses from this model's FE analysis at the free end of the cracked cantilever beam under moving load was plotted against the analytical results in the paper by Lin and Chang [25], and it can be clearly seen that the two curves trace one another closely almost everywhere, shown in Figure 5, indicating the accuracy of the proposed model in this experimental range.

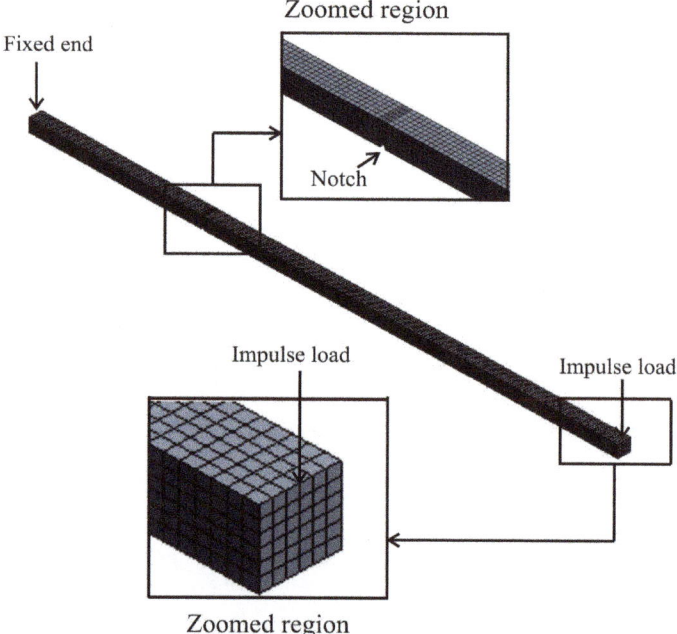

Figure 4. Beam with meshes assigned for FEA analysis.

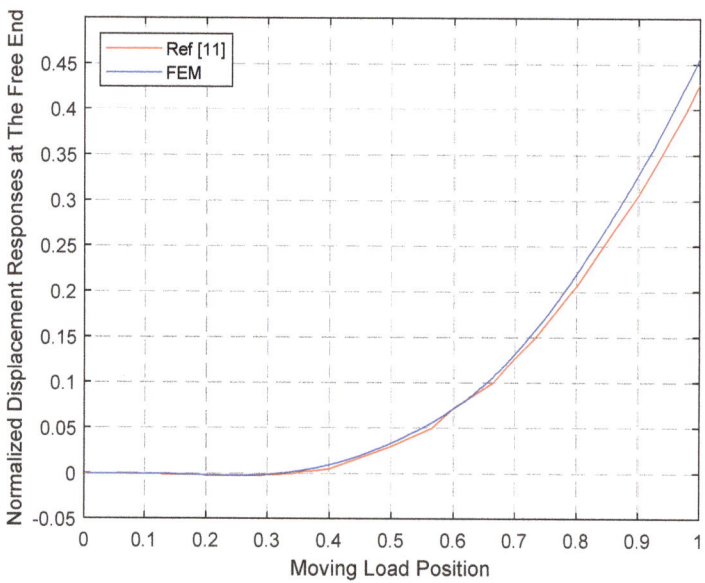

Figure 5. Force deflection response of the modeled cracked beam against analytical response.

2.5. Evaluation of the Proposed Method

The location of a crack found by the proposed method was evaluated against that detected by the mode shape curvature method in the paper by Pendy et al. [5]. In the evaluation of the proposed method, calculation of displacement at each point in and on the beam in response to moving load was performed by ANSYS. The displacement calculation method, selected from a menu in ANSYS, was the Newmark's integration scheme, with a fixed time increment of 0.00001 s.

The assigned 3D model and FEA settings were the following.

1. Element type Selection: SOLID 186 Hex 20 node brick elements.
2. FE Model Creation (Meshing): Mesh generation is a process of dividing the structure continuum into a number of discrete parts or finite elements. In this study, the uncracked beam was assigned a total number of FE elements of 4620 and the total number of nodes of 24,768. For the cracked beam, the total number of elements was the same, at 4620, but the number of nodes were higher, at 2.4803, because a higher number of meshes were assigned at the crack region to model it more precisely. The meshes were constructed using 232 rows of elements along the length of the beam, six rows of elements across the width, and four rows of elements through its depth or thickness. The maximum size of a mesh was 0.25 mm. The boundary condition for one end of the beam was that it was fixed to a location in the environment, while the boundary condition for the other end was that it could move freely in response to various forces acting on it. The illustration of the meshed FE model of the cracked beam is shown in Figure 4.
3. Assigned material properties: assigned Young's modulus and Poisson's ratio are listed in Table 1.
4. Applied loads: concentrated load (F) acting on a point of contact on the surface of the cantilever beam, moving from the left end to the right end of the beam at a speed of 30.9 m/s. This speed was 0.6 time critical speed V_{crit} as defined by Lin and Chang [25]. The tested magnitudes of the moving load were 70 N, 80 N, 90 N, and 100 N. Five replications were conducted for each magnitude. The goal was to determine which moving load magnitude would provide the most distinguishable peak in a graph of central difference approximation of displacement versus location coordinates (detailed in Section 2.1). The way that this proposed method calculated the displacement of a point in and on a beam depended only on the concentrated load, F, exerting itself on the numbered nodes—1, 2, 3, . . . , m—as shown in Figure 1.

For the curvature mode shape method, the magnitude of the first displacement mode shape was used to calculate the curvature mode shape with the Modal Module in ANSYS. The assigned settings for the beam were the same as in the evaluation of the proposed method.

We assigned a crack to be at one-third, the middle, and two-thirds of beam length and identified their locations. Then, we did the same but using the proposed method and compared the results. Figures 6–8 show plots (with 10^{-4} scaling) of three crack locations identified by the proposed method (for which a moving load of 70 N was applied) overlayed on plots of the three crack locations identified by Pendy's curvature mode shape method.

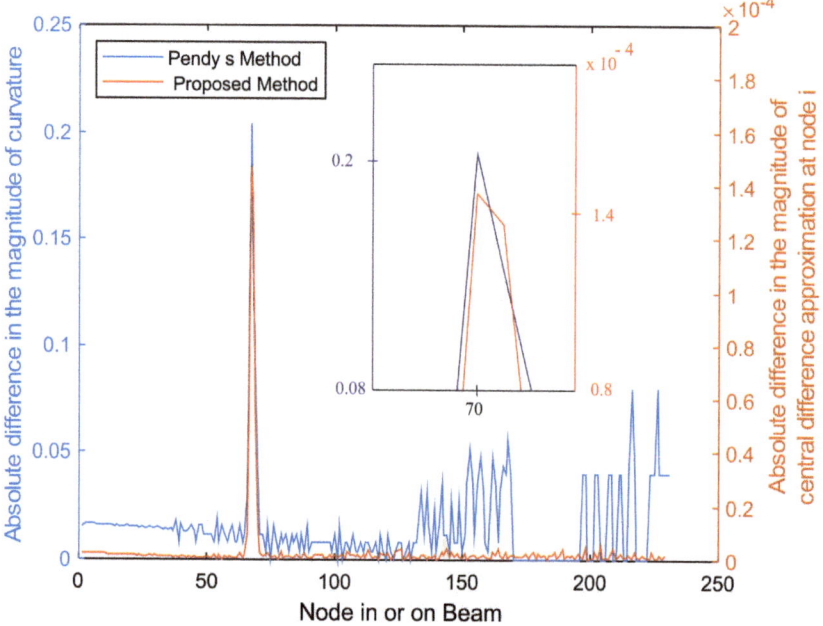

Figure 6. Crack at node 70 (at one-third of beam length).

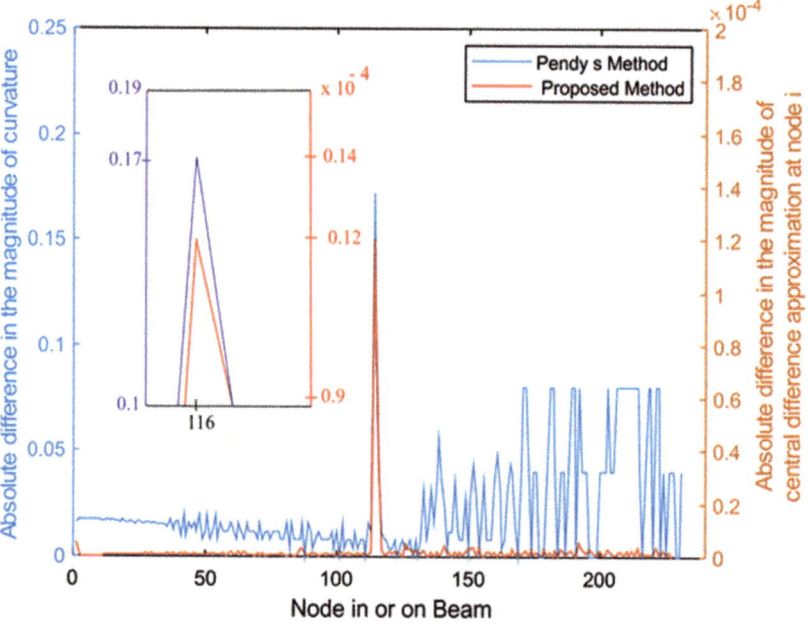

Figure 7. Crack at node 116 (at the middle of the beam).

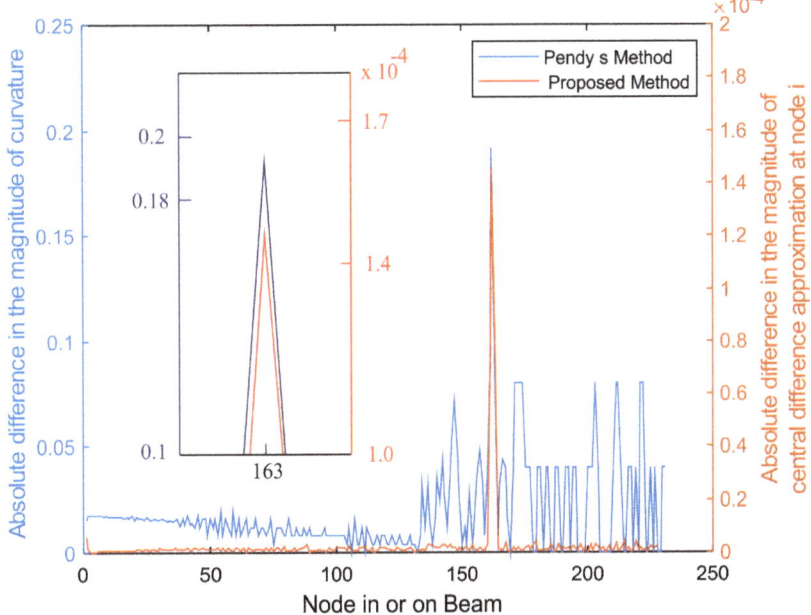

Figure 8. Crack at node 163 (at two-thirds of beam length).

3. Numerical Results and Discussions

This section discusses five main points: Verification of forced vibration behavior of a cracked beam calculated by FEA against that of Hai-Ping Lin and Shun-Chang Chang's exact method [25]; Inability of average dynamic deflections of cantilever beam under moving load to detect a crack by themselves; Dependence of difference in central difference approximation of displacement of cracked beam and uncracked beam on the magnitude of moving load; Comparison of crack locations identified by the proposed method and the mode shape method of Pendy et al., [5]; and Advantage of FEA in comparison with Lin and Chang's analytical method.

3.1. Verification of Forced Vibrational Behavior Using FEA

To check the validity of the forced vibration behavior of a cracked cantilever beam under moving load calculated by FEA, the results from FEA analysis were verified against such behavior calculated by Hai-Ping Lin and Shun-Chang Chang's exact method. The graphs of the results from both methods were overlayed on top of the other in Figure 8. It can be clearly observed that in the range of position from 0–0.7 beam length, the two graphs almost coincided, while they did not deviate more than 5% in the range of position from 0.7–0.9 beam length. Therefore, it can be concluded that the beam vibrational behavior was modeled sufficiently precisely for crack identification by the proposed crack-identification method.

3.2. Average Dynamic Deflection Results

Average dynamic deflection was evaluated by Equation (6). The calculated results are plotted against the position of the beam under four magnitudes of moving load in Figures 9–11, where a crack was positioned at x/l = 0.3, 0.5, and 0.7, respectively. It can be observed that all four curves in each figure are smooth curves that do not show a peak or trough at any position, i.e., a crack cannot be identified directly from these curves under any magnitudes of moving load. To conclude, average dynamic deflection data alone cannot be used to identify a crack. However, the higher the moving load, the higher the magnitude of average deflection or average displacement.

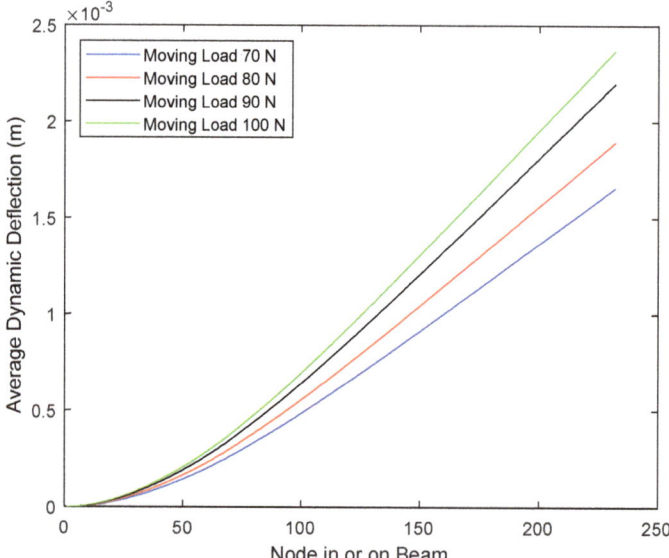

Figure 9. Average Dynamic Deflection of Cracked Beam (crack at $x/l = 0.30$) under moving load.

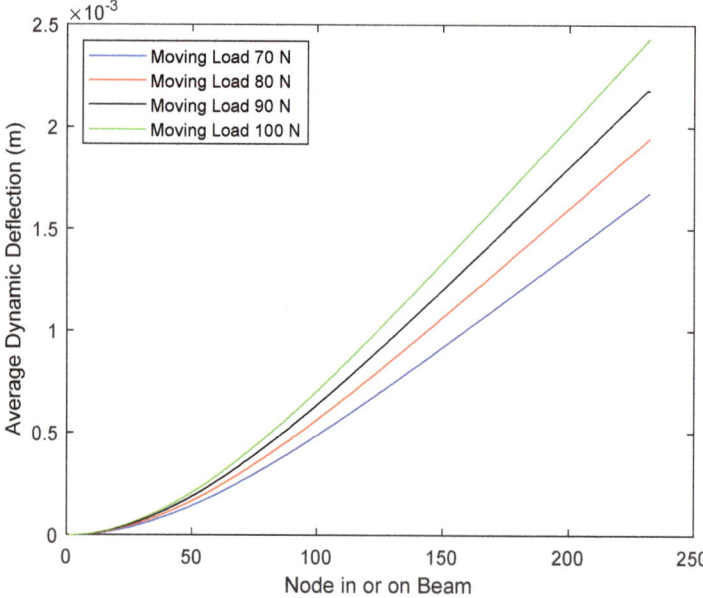

Figure 10. Average Dynamic Deflection of Cracked Beam (crack at $x/l = 0.50$) under moving load.

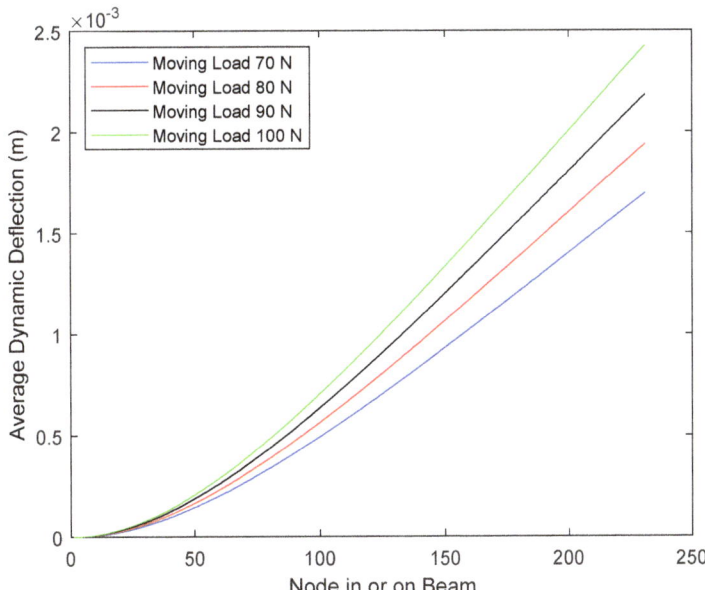

Figure 11. Average Dynamic Deflection of Cracked Beam (crack at $x/l = 0.70$) under moving load.

It can be observed in Figures 12 and 13 that the average dynamic deflection of the uncracked beam was the lowest. However, it cannot be used to indicate the crack position in a cracked beam.

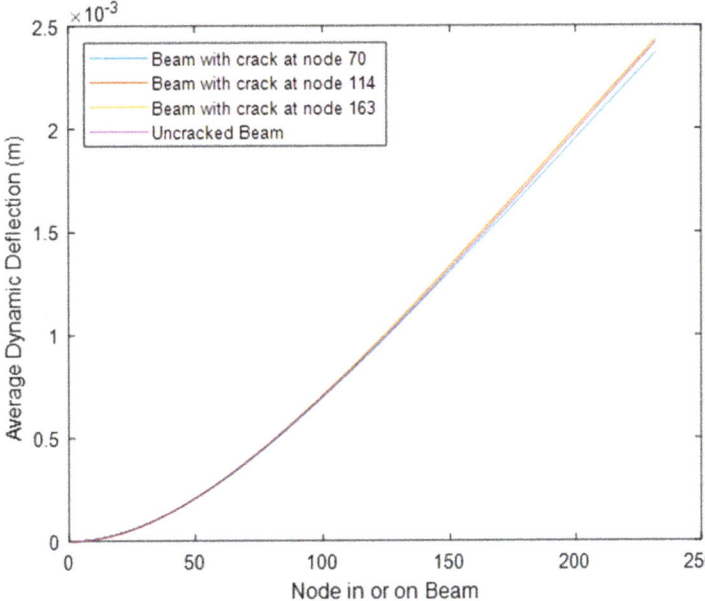

Figure 12. Average dynamic deflection of cracked and uncracked beam under a 100 N moving load.

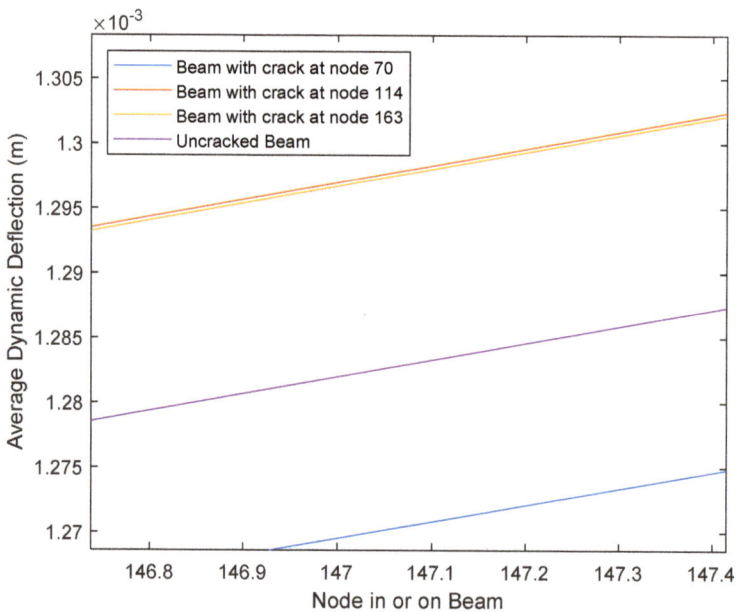

Figure 13. Blown-up view showing clearly the non-overlapping nature of the curves in Figure 12.

3.3. Dependence of Difference in Central Difference Approximation on the Magnitude of Moving Load

Even though the graph of average dynamic deflection or average vertical displacement of a vibrational cracked beam under moving load could not show the location of a crack, a plot of the difference in central difference approximation of displacement of the cracked beam and an uncracked beam under moving mass did show a sharp peak at the crack location on the cantilever beam due to the change in stiffness of the beam at the crack.

Figures 14–16 illustrate the results obtained from the absolute difference in the central difference approximation of displacement at a node of a cracked beam and an uncracked beam, evaluated by Equation (3), at each of three wedge-shaped notch locations: a notch (or a crack) at $x_1/l = 0.3$; a notch at mid-span of the beam; and a notch at $x_1/l = 0.75$. The notch depth was 30% of the beam thickness. It can be observed that for all notch locations, varying the moving load from 70 N to 100 N still provided clearly distinguishable peaks at the same notch location, though a load of 100 N provided the highest peak because it caused more deflection or displacement.

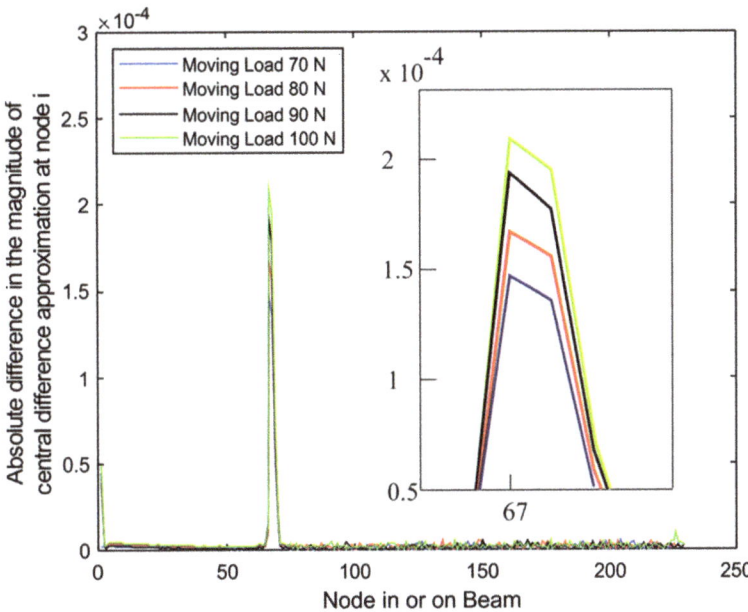

Figure 14. Absolute difference between the central difference approximation of displacement of the uncracked and the cracked (Node 70) cantilever beam subjected to moving load.

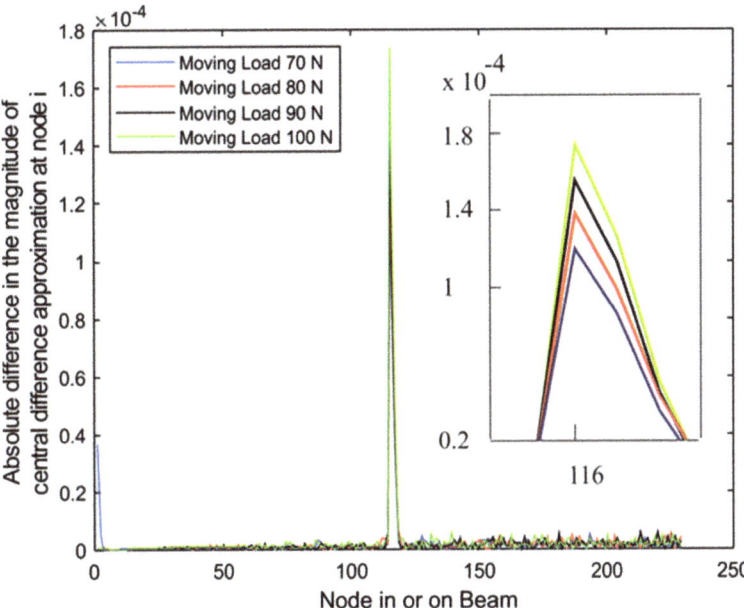

Figure 15. Absolute difference between the central difference approximation of displacement of the uncracked and the cracked (Node 116) cantilever beam subjected moving load.

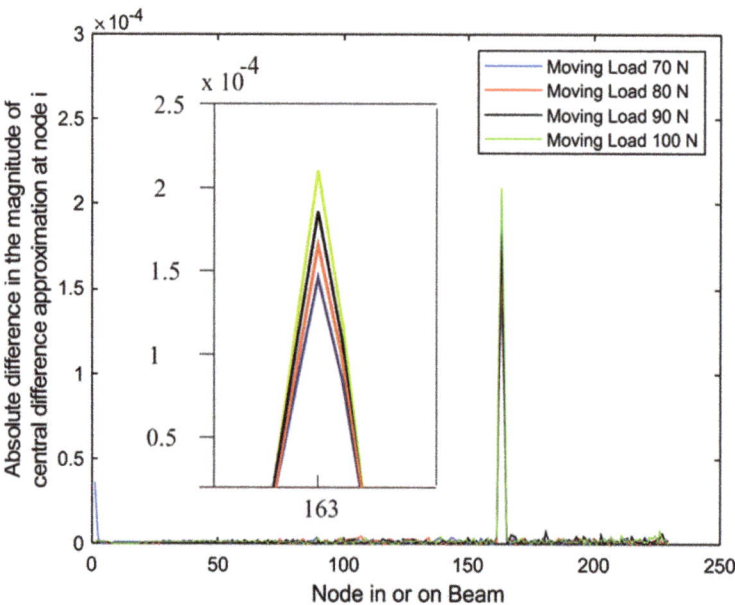

Figure 16. Absolute difference between the central difference approximation of displacement of the uncracked and the cracked (Node 163) cantilever beam subjected moving load.

3.4. Comparison of Crack Locations Identified by the Proposed Method and Pendy's Mode Shape Method

Figures 6–8 show plots (with 10^{-4} scaling) of three crack locations identified by the proposed method (for which a moving load of 70 N was applied) overlayed on plots of the three crack locations identified by Pendy's curvature mode shape method. It can be clearly seen that the locations were all corresponding to the exact same locations, showing that the proposed method could be used as an alternative to Pendy's analytical method without any significant errors for identifying a crack in or on a cantilever beam.

3.5. Advantage of FEA in Comparison with Hai-Ping Lin and Shun-Chang Chang's Analytical Method

The operation of FEA was performed to evaluate the efficiency of the proposed method, as an alternative to conducting an experiment or using an analytical method to simulate vibration of the cracked beam and the uncracked beam.

At present, some kinds of vibration simulation of a cracked beam can be conducted by analytical method [25], mostly analyzed with Fourier series. However, analytical solutions to different sets of governing equation and boundary conditions may be too complex to solve or even unsolvable for some specific applications. Engineers, however, want a fast and familiar way to tackle a problem, instead of trying to find an analytical solution to the problem and spend a lot of time and effort to code it in an efficient computational programming language that they are not familiar with in order to finish a project in a sufficiently short time. The bread-and-butter ready-made application program—SolidWorks, Ansys FEA, and Matlab—that all engineers are familiar with are the tools that they can use to develop a project much faster and easier than finding and coding an analytical solution of the same problem. For instance, engineers only need to construct a 3D model of the object and input it directly into Ansys without having to write a procedure to import a 3D file into a manually coded program; then, they only need to assign boundary conditions, there is no need to write code for them, and they can assign the moving load from a menu and run the simulation, all with a few clicks of a mouse. The amount of time to complete a project, most likely, approximates the amount of time for finding an analytical solution and manually coding a program to run the simulation (for engineers that are quite comfortable with the programming language). As shown in our work, the result for crack identification was virtually identical, and the economy in time and effort was real.

4. Conclusions

In the present work, a crack-identification method by FEA is presented based on an assumption of a transverse surface crack, extending uniformly along the width of a cantilever beam. In a numerical study, forced responses of an uncracked and a cracked cantilever beam under a moving load were simulated by transient analysis using the Ansys FEA software. Various notch locations for a crack were investigated. A time domain procedure relying on measured time response was presented. Displacement at each point on the beam was obtained from probe deformation menu for transient analysis in Ansys software.

For forced responses of cantilever beams under a moving load, a numerical procedure based on curvature mode shape was developed. Forced responses of cracked beams were investigated with varying moving load. Locations of cracks were indicated by peaks of absolute changes in central difference approximation of displacement plot. The effectiveness of the proposed damage-identification scheme was positively verified by, first, its ability to find the exact same location of the simulated crack as the exact solution method has found, and second, by its simpler one-vibrational-mode procedure than the multi-vibrational-mode that the curvature mode shape method of Pendy [5], quite often, required. Therefore, it has a great application potential with the following advantage: only the deflection parameters of the beam are needed to perform damage localization, and the height of curvature peaks can be varied by adjusting the weight of the moving load. The method can be regarded as a relatively simple, low cost, and effective tool for nondestructive testing (NDT) that does not require sophisticated equipment.

Since virtually every engineer is familiar with Ansys FEA but much less familiar with coding an exact solution (when derivable) in a programming language, the advantage of this method is that it can provide the location of a crack as accurately as an actual experiment or an analytical method can, while taking much less time and effort to develop and implement. With very little amount of trial and error, it can be expected that this method will also accurately identify crack location in a more complex structure with difficult-to-code boundary conditions.

5. Future Work and Limitations of the Proposed Method

This study was a first exploration and development of a crack-identification method, so it is not extensive: (i) Only Euler beam type cantilever was investigated; (ii) the method assumes a single damage location; and (iii) No noise was introduced to the simulated data to test its robustness.

In the near future, we will test the method on various practical structures as case studies. The method is a promising step towards establishing a practical and reliable piping health monitoring procedure, where the location of a potential crack (such as a weld line) is known beforehand.

Author Contributions: Conceptualization, N.K. and T.J.; methodology, N.K. and T.J.; software, N.K. and T.J.; validation, N.K. and T.J.; formal analysis, N.K. and T.J.; investigation, N.K., T.J., S.K. and C.T.; data curation, N.K. and T.J.; writing—original draft preparation, N.K., T.J. and S.K.; writing—review and editing N.K., T.J. and S.K.; visualization, N.K., T.J., S.K. and C.T.; supervision, T.J.; project administration, T.J.; funding acquisition, T.J.; All authors have read and agreed to the published version of the manuscript.

Funding: This research received no external funding.

Institutional Review Board Statement: Not applicable.

Informed Consent Statement: Not applicable.

Data Availability Statement: The data and materials in this paper are available.

Acknowledgments: This work was supported by Thammasat University Research Unit in Structural and Foundation Engineering.

Conflicts of Interest: The authors declare no conflict of interest.

References

1. Concli, F.; Kolios, A. Preliminary Evaluation of the Influence of Surface and Tooth Root Damage on the Stress and Strain State of a Planetary Gearbox: An Innovative Hybrid Numerical—Analytical Approach for Further Development of Structural Health Monitoring Models. *Computation* **2021**, *9*, 38. [CrossRef]
2. Cheng, Q.; Ruan, X.; Wang, Y.; Chen, Z. Serious damage localization of continuous girder bridge by support reaction influence lines. *Buildings* **2022**, *12*, 182. [CrossRef]
3. Duvnjak, I.; Klepo, I.; Serdar, M.; Damjanović, D. Damage Assessment of reinforced concrete elements due to corrosion effect using dynamic parameters: A review. *Buildings* **2021**, *11*, 425. [CrossRef]
4. Dimarogonas, A.D. *Vibration for Engineers*, 2nd ed.; Prentice Hall: Hoboken, NJ, USA, 1996.
5. Pandey, A.; Biswas, M.; Samman, M. Damage detection from changes in curvature mode shapes. *J. Sound Vib.* **1991**, *145*, 321–332. [CrossRef]
6. Sahin, M.; Shenoi, R.A. Quantification and localisation of damage in beam-like structures by using artificial neural networks with experimental validation. *Eng. Struct.* **2003**, *25*, 1785–1802. [CrossRef]
7. Ratcliffe, C.P.; Bagaria, W.J. Vibration technique for locating delamination in a composite beam. *AIAA J.* **1998**, *36*, 1074–1077. [CrossRef]
8. Hamey, C.S.; Lestari, W.; Qiao, P.; Song, G. Experimental damage identification of carbon/epoxy composite beams using curvature mode shapes. *Struct. Health Monit.* **2004**, *3*, 333–353. [CrossRef]
9. Qiao, P.H.; Lu, K.; Lestari, W.; Wang, J. Curvature mode shape-based damage detection in composite laminated plates. *Compos. Struct.* **2007**, *80*, 409–428. [CrossRef]
10. Chandrashekhar, M.; Ganguli, R. Damage assessment of structures with uncertainty by using mode-shape curvatures and fuzzy logic. *J. Sound Vib.* **2009**, *326*, 939–957. [CrossRef]
11. Frans, R.; Arfiadi, Y.; Parung, H. Comparative study of mode shapes curvature and damage locating vector methods for damage detection of structures. *Procedia Eng.* **2017**, *171*, 1263–1271. [CrossRef]
12. Lin, Y.Z.; Zhao, Z.; Nie, Z.H.; Ma, H.W. Data-driven damage detection for beam-like structures under moving loads using quasi-static responses. *Front. Artif. Intell. Appl.* **2016**, *281*, 403–411.
13. Kordestani, H.; Xiang, Y.Q.; Ye, X.W. Output-only damage detection of steel beam using moving average filter. *Shock. Vib.* **2018**, *2018*, 2067680. [CrossRef]
14. Kordestani, H.; Xiang, Y.Q.; Ye, X.W.; Jia, Y.K. Application of the random decrement technique in damage detection under moving load. *Appl. Sci.* **2018**, *8*, 753. [CrossRef]
15. Senjuntichai, T.; Keawsawasvong, S.; Yooyao, B. Exact stiffness method for multi-layered saturated soils under moving dynamic loads. *J. GeoEng.* **2020**, *15*, 159–171.
16. Liu, Y.Z.; Yi, T.H.; Yang, D.H.; Li, H.N. Damage location of beam railway bridges using rotation responses under moving train loads. *J. Perform. Constr. Facil.* **2021**, *35*, 04021096. [CrossRef]
17. Chouiyakh, H.; Azrar, L.; Alnefaie, K.; Akourri, O. Vibration and multi-crack identification of Timoshenko beams under moving mass using the differential quadrature method. *Int. J. Mech. Sci.* **2017**, *120*, 1–11. [CrossRef]
18. Roveri, N.; Carcaterra, A. Damage detection in structures under traveling loads by Hilbert-Huang transform. *Mech. Syst. Signal Process.* **2012**, *28*, 128–144. [CrossRef]
19. Ghannadiasl, A.; Khodapanah Ajirlou, S. Dynamic analysis of multiple cracked Timoshenko beam under moving load—Analytical method. *J. Vib. Control* **2022**, *28*, 379–395. [CrossRef]
20. Kourehli, S.S.; Ghadimi, S.; Ghadimi, R. Vibration analysis and identification of breathing cracks in beams subjected to single or multiple moving mass using online sequential extreme learning machine. *Inverse Probl. Sci. Eng.* **2019**, *27*, 1057–1080. [CrossRef]
21. Zhu, X.; Cao, M.; Ostachowicz, W.; Xu, W. Damage identification in bridges by processing dynamic responses to moving loads: Features and evaluation. *Sensors* **2019**, *19*, 463. [CrossRef]
22. Qiao, G.; Rahmatalla, S. Dynamics of Euler-Bernoulli beams with unknown viscoelastic boundary conditions under a moving load. *J. Sound Vib.* **2021**, *491*, 115771. [CrossRef]
23. Asnaashari, E.; Sinha, J.K. Crack detection in structures using deviation from normal distribution of measured vibration responses. *J. Sound Vib.* **2014**, *333*, 4139–4151. [CrossRef]
24. Lee, H.-H. *Finite Element Simulations with ANSYS Workbench 18*; SDC Publications: Mission, KS, USA, 2018.
25. Lin, H.-P.; Chang, S.-C. Forced responses of cracked cantilever beams subjected to a concentrated moving load. *Int. J. Mech. Sci.* **2006**, *48*, 1456–1463. [CrossRef]
26. Ansys. *Finite Element Computer Software for Nonlinear Structural Analysis, Version 15.0*; Ansys Inc.: Canonsburg, PA, USA, 2013.

MDPI
St. Alban-Anlage 66
4052 Basel
Switzerland
www.mdpi.com

Computation Editorial Office
E-mail: computation@mdpi.com
www.mdpi.com/journal/computation

Disclaimer/Publisher's Note: The statements, opinions and data contained in all publications are solely those of the individual author(s) and contributor(s) and not of MDPI and/or the editor(s). MDPI and/or the editor(s) disclaim responsibility for any injury to people or property resulting from any ideas, methods, instructions or products referred to in the content.

www.ingramcontent.com/pod-product-compliance
Lightning Source LLC
LaVergne TN
LVHW070652100526
838202LV00013B/945